Advance praise fo[illegible]

"A superior account of how astrono[illegible] that they knew almost nothing about 96 perce[illegible]....  Panek delivers vivid sketches of scientists, luc[illegible] explanations of their work, and revealing descriptions of the often stormy rivalry that led to this scientific revolution."

***Kirkus Reviews***, starred review

"It's the biggest mystery of all: why is the universe expanding at an accelerated rate? It baffled Einstein, and it now obsesses a cadre of fascinating cosmologists. By brilliantly capturing their passions and pursuits, Richard Panek has made this cosmic quest exciting and understandable."

**Walter Isaacson**, author of *Einstein: His Life and Universe*

"Somebody needed to tell this story – of all that is dark and mysterious in the cosmos. Science writer Richard Panek has risen to that task. He artfully guides you through the quirky discoveries that established the existence of dark matter and dark energy. But along the way, you also get to meet the quirky cosmologists who got us there."

**Neil deGrasse Tyson**, Frederick P. Rose Director of the Hayden Planetarium at the American Museum of Natural History

"*The 4 Percent Universe* is a reliable and readable account of how scientists discovered – and are struggling to come to grips with – the astounding fact that most of the observable universe has not yet even been observed, much less understood. It has the further merit of relating how scientists arrive at their findings, rather than simply presenting their theories as objects of admiration or adoration. Highly recommended."

**Timothy Ferris**, author of *Coming of Age in the Milky Way* and Emeritus Professor at the University of California-Berkeley

"Modern cosmology tackles some of the biggest questions we have about the nature of the cosmos. In *The 4 Percent Universe,* Richard Panek brings this quest down to a human scale. The rivalries, the surprises, and the excitement are brought vividly to life. People are a very tiny percentage of the universe, but we remain the most interesting part."

**Sean Carroll**, author of *From Eternity to Here*

"A lively and well-researched account of the personalities and ambitions of modern scientists."

**Alan Lightman**, author of *Einstein's Dreams*

"Richard Panek has written a contemporary adventure story of modern-day explorers who venture forth into the universe not by ships, but by telescopes and satellites. Like adventure stories of old, there are visionaries, heroes, patrons, and, perhaps, a few pirates. A riveting book."

**Lee Smolin**, author of *The Trouble with Physics*

"A compelling story of research at the cutting edge of science, with all the personalities and politics that the textbooks leave out."

**Chad Orzel**, author of *How to Teach Quantum Physics to Your Dog*

"Richard Panek turns astronomers and physicists into real (and sometimes likeable) characters. You can feel the tension as two rival groups race to discover the fate of the universe. We see scientists as real people, warts and all. Panek transforms potentially baffling science into a tense story."

**Brian Clegg**, author of *Before the Big Bang* and *Armageddon Science*

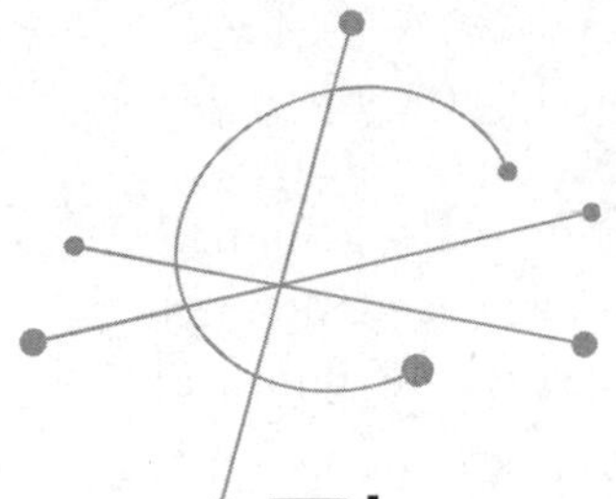

# The 4 Percent Universe

## DARK MATTER, DARK ENERGY, AND THE RACE TO DISCOVER THE REST OF REALITY

Richard Panek

ONEWORLD
OXFORD

A Oneworld Book

First published in Great Britain and the Commonwealth by
Oneworld Publications 2011

First published in the USA by Houghton Mifflin Harcourt
Publishing Company, 2011


A CIP record for this title is available
from the British Library

ISBN 978-1-85168-821-0

Cover design by Richard Green
Printed and bound in Great Britain by Page Bros

Oneworld Publications
185 Banbury Road, Oxford, OX2 7AR, England



Some passages in this book appeared, usually in different form, in *Discover,* the *New York Times Magazine, Sky & Telescope,* and *Smithsonian.*
Portions of chapter 11 are based on work supported by the National Science Foundation under Grant No. 0739893. Any opinions, findings, and conclusions or recommendations expressed in this material are those of the author and do not necessarily reflect the views of the National Science Foundation.

*For Meg, with love*

"I know," said Nick.

"You don't know," said his father.

—*Ernest Hemingway*

# Contents

# Acknowledgements

THE AUTHOR EXPRESSES deep appreciation to Amanda Cook for her extraordinary editorial guidance as well as her genuine passion for the dark side of the universe; Henry Dunow, who, with his usual wisdom, made the match of editor and author; Katya Rice for her expert eye; Katherine Bouton for taking a chance on science and assigning an article on this subject; the John Simon Guggenheim Memorial Foundation, the National Science Foundation's Antarctic Artists and Writers Program, and the New York Foundation for the Arts for their generous and essential support; and Gabriel and Charlie (who claim to know what dark matter and dark energy are but refuse to tell their father, who loves them anyway).

# Prologue

THE TIME HAD come to look inside the box. On 5 November 2009, scientists at sixteen institutions around the world took their seats before their computer screens and waited for the show to begin: two software programs being run by two graduate students — one at the University of Minnesota, the other at the California Institute of Technology — simultaneously. For fifteen minutes the two scripts would sort through data that had been collecting far underground in a long-abandoned iron mine in northern Minnesota. Over the past year, thirty ultrasensitive detectors — deep-freeze cavities the size of refrigerators, shielded from stray cosmic rays by half a mile of bedrock and snug blankets of lead, their interiors cooled almost to absolute zero, each interior harboring a heart of germanium atoms — had been looking for a particular piece of the universe. The data from that search had sped from the detectors to offsite computers, where, following the protocol of a blind analysis, it remained in a "box," out of sight. Just after 9 A.M., the "unblinding party" began.

Jodi Cooley watched on the screen in her office at Southern Methodist University. As the coordinator of data analysis for the experiment, she had made sure that researchers wrote the two scripts separately using two independent approaches, so as to further ensure against bias. She had also arranged for all the collaborators on the project — physicists at Stanford, Berkeley, Brown; in Florida, Ohio Switzerland — to be sitting at their computers at the same time. Together they would watch the evidence as it popped up on their screens, one plot per detector, two versions of each plot.

After a few moments, plots began appearing. Nothing. Nothing. Nothing.

Then, three or four minutes into the run, a detection appeared—on the same plots in both programs. A dot on a graph. A dot within a narrow, desirable band. A band where all the other dots weren't falling.

A few minutes later another pair of dots on another pair of plots appeared within the same narrow band.

And a few minutes later the programs had run their course. That was it, then. Two detections.

"Wow," Cooley thought.

*Wow*, as in: They had actually seen something, when they had expected to get the same result as the previous peek inside a "box" of different data nearly two years earlier—nothing.

*Wow,* as in: If you're going to get detections, two is a frustrating number—statistically tantalizing but not sufficient to claim a discovery.

But mostly *Wow,* as in: They might have gotten the first glimpse of dark matter—a piece of our universe that until recently we hadn't even known to look for, because until recently we hadn't realized that our universe was almost entirely missing.

It wouldn't be the first time that the vast majority of the universe turned out to be hidden to us. In 1610 Galileo announced to the world that by observing the heavens through a new instrument—what we would call a telescope—he had discovered that the universe consists of more than meets the eye. The five hundred copies of the pamphlet announcing his results sold out immediately; when a package containing a copy arrived in Florence, a crowd quickly gathered around the recipient and demanded to hear every word. For as long as members of our species had been lying on our backs, looking up at the night sky, we had assumed that what we saw was all there was. But then Galileo found mountains on the Moon, satellites of Jupiter, hundreds of stars. Suddenly we had a new universe to explore, one to which astronomers would add, over the next four centuries, new moons around other planets, new planets around our Sun, hundreds of planets around other stars, a hundred billion stars in our galaxy, hundreds of billions of galaxies beyond our own.

By the first decade of the twenty-first century, however, astronomers had concluded that even this extravagant census of the universe might be as out-of-date as the five-planet cosmos that Galileo inherited from the ancients. The new universe consists of only a minuscule fraction of what we had always assumed it did—the material that makes up you and me and my laptop and all those moons and planets and stars and galaxies. The rest—the overwhelming majority of the universe—is . . . who knows?

"Dark," cosmologists call it, in what could go down in history as the ultimate semantic surrender. This is not "dark" as in distant or invisible. This is not "dark" as in black holes or deep space. This is "dark" as in unknown for now, and possibly forever: 23 percent something mysterious that they call dark matter, 73 percent something even more mysterious that they call dark energy. Which leaves only 4 percent the stuff of us. As one theorist likes to say at public lectures, "We're just a bit of pollution." Get rid of us and of everything else we've ever thought of as the universe, and very little would change. "We're completely irrelevant," he adds, cheerfully.

All well and good. Astronomy is full of homo sapiens–humbling insights. But these lessons in insignificance had always been at least somewhat ameliorated by a deeper understanding of the universe. The more we could observe, the more we would know. But what about the less we could observe? What happens to our understanding of the universe then? What currently unimaginable repercussions would this limitation, and our ability to overcome it or not, have for our laws of physics and our philosophy—our twin frames of reference for our relationship to the universe?

Astronomers are finding out. The "ultimate Copernican revolution," as they often call it, is taking place right now. It's happening in underground mines, where ultrasensitive detectors wait for the ping of a hypothetical particle that might already have arrived or might never come, and it's happening in ivory towers, where coffee-break conversations conjure multiverses out of espresso steam. It's happening at the South Pole, where telescopes monitor the relic radiation from the Big Bang; in Stockholm, where Nobelists have already begun to receive recognition for their encounters with the dark side; on the laptops of postdocs around the world, as they observe the real-time self-annihilations of stars, billions of light-years distant, from the comfort of a living room couch. It's happening in healthy collaborations and, the universe being the intrinsically Darwinian place it is, in career-threatening competitions.

The astronomers who have found themselves leading this revolution didn't set out to do so. Like Galileo, they had no reason to expect that they would discover new phenomena. They weren't looking for dark matter. They weren't looking for dark energy. And when they found the evidence for dark matter and dark energy, they didn't believe it. But as more and better evidence accumulated, they and their peers reached a consensus that the universe we thought we knew, for as long as civilization had been looking at the night sky, is only a shadow of what's out there. That we have been blind to the actual universe because it consists of less than meets the eye. And that *that* universe is our universe — one we are only beginning to explore.

It's 1610 all over again.

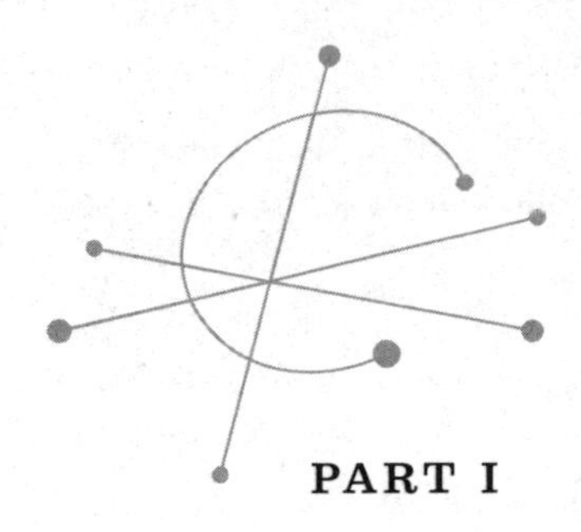

PART I

# More Than Meets the Eye

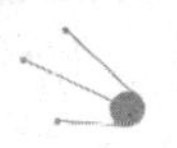

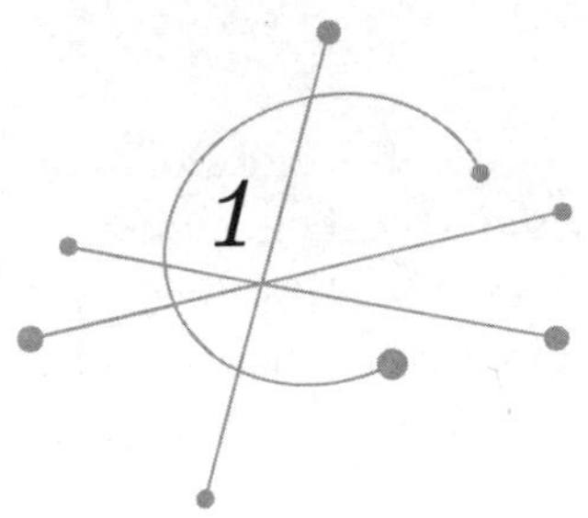


# Let There Be Light

IN THE BEGINNING—which is to say, 1965—the universe was simple. It came into being one noontime early that year over the course of a telephone conversation. Jim Peebles was sitting in the office of his mentor and frequent collaborator, the Princeton physicist Robert Dicke, along with two other colleagues. The phone rang; Dicke took the call. Dicke helped run a research firm on the side, and he himself held dozens of patents. During these weekly lunches in his office, he sometimes got phone calls that were full of esoteric and technical vocabulary that Peebles didn't recognize. This call, though, contained esoteric and technical vocabulary that Peebles knew intimately—concepts the four physicists had been discussing that very afternoon. Cold load, for instance: a device that would help calibrate the horn antenna—another term Peebles overheard—that they would be using to try to detect a specific signal from space. The three physicists grew quiet and looked at Dicke. Dicke thanked the caller and hung up, then turned to his colleagues and said, "Well, boys, we've been scooped."

The caller was an astronomer at the Bell Telephone Laboratories who had collected some curious data but had no idea what it meant. Peebles and Dicke had developed a curious idea but had no data to support it. The other two physicists at lunch had been building an antenna to detect a signal that would offer support for the curious idea, but now, Dicke said, a pair of astronomers at Bell Labs had probably found it first—without knowing what they'd done.

The mood in Dicke's office was not one of deflation or disappointment. If the four of them had in fact been scooped, they had also been vindicated. If the caller was right, then they too were right, or at least heading in a potentially profitable scientific direction. If nothing else, they could take some comfort from the possibility that they were the first persons in the history of the world to understand the history of the universe.

But before reaching any conclusions, they would have to check the data for themselves. Dicke and two of the other Princeton physicists soon drove the thirty miles to Holmdel Township, New Jersey, home of the Bell Labs research centre. The Bell Labs astronomers—Arno Penzias, who had placed the call to Dicke, and his collaborator Robert Wilson—took them to see the antenna. It was a horn-shaped instrument, as big as a freight carriage, by the side of a private road at the top of Crawford Hill, the highest point for miles around. After the five of them had squeezed into the control cab, their elbows brushing vacuum tubes and instrument panels, the Bell Labs astronomers explained the physics to the physicists.

Bell Labs had built the antenna in 1960 to receive coast-to-coast signals bouncing off the Echo communications satellite, a highly reflective balloon 30 metres in diameter. When the Echo mission ended, the antenna was used on the Telstar satellite. When that mission ended, Penzias and Wilson appropriated the antenna to study radio waves from the fringes of our Milky Way galaxy. The measurements would have to be much more sensitive than they were for Echo, so Penzias had built a cold load, an instrument that emitted a specific signal that he and Wilson could compare with measurements from the antenna to make sure it wasn't detecting any excess noise. And the cold load worked, just not in the way they'd hoped. Aside from the unavoidable rattling of electrons in the atmosphere and within the instrument itself, Penzias and Wilson were left with a persistent, inexplicable hiss.

For much of the past year they had been trying to determine the source of the noise. They pointed the antenna at New York City, less than fifty miles away. The radio static was negligible. They pointed the antenna at every other location on the horizon. The same. They checked the signal from the stars to see whether it differed from what they'd already factored into their calculations. Nope. The phases of the Moon? Temperature changes in the atmosphere over the course of a year? No and no. That spring they had turned their attention back to the antenna itself. They put tape over the aluminium rivets in the antenna—nothing—and took apart the throat of the horn and put it back together again—nothing—and even scraped away the droppings from a pair of pigeons that had taken up residence within the horn. (They caught the pigeons and posted them to the Bell Labs site in Whippany, New Jersey, more than forty miles away; the birds turned out to be homing pigeons and were back in the horn within days.) Still nothing—nothing but the noise.

The five scientists repaired to a conference room on Crawford Hill, and now the physicists explained the astronomy to the astronomers. Dicke started writing on a blackboard. If the Big Bang interpretation of the history of the universe was correct, Dicke said, then the cosmos emerged in an unfathomably condensed, obscenely hot explosion of energy. Everything that would ever be in the universe was there then, rushing outwards on a shock wave of space itself, and continuing to rush outwards until it evolved into the universe we see today. And as the universe expanded, it cooled. One member of the Princeton collaboration—Jim Peebles, the colleague who wasn't present—had calculated what that initial level of energy would have been, and then he had calculated what the current level of energy, after billions of years of expansion and cooling, should be. That remnant energy—assuming it existed; assuming the Big Bang theory was right—would be measurable. And now, apparently, Penzias and Wilson *had* measured it. Their antenna was picking up an echo, all right, but this time the source wasn't a radio broadcast from the West Coast. It was the birth of the universe.

Penzias and Wilson listened politely. Dicke himself didn't entirely believe what he was saying—not yet. He and the two other Princeton physicists satisfied themselves that Penzias and Wilson had run a clean experiment, then drove back to Princeton and told Peebles what they had learned. Peebles didn't entirely believe what he was hearing, either. He was cautious; but then, he was always cautious. The four collaborators agreed that scientific results require corroboration, a second opinion—in this case, their own. They would finish constructing their antenna on the roof of Princeton's Guyot Hall and see if it got the same reading as the Bell Labs antenna. Even if it did, they knew they would *still* have to proceed with caution. It's not often, after all, that you get to discover a new vision of the universe.

The American writer Flannery O'Connor once said that every story has "a beginning, a middle, and an end, though not necessarily in that order." By the 1960s, scientists wanting to tell the story of the universe — cosmologists, by definition — could proceed under the assumption that they were in possession of the middle of the narrative. They had the latest version of one of civilization's most enduring characters, the universe — in this case, an expanding one. Now they could ask themselves: How did Our Hero get here?

The capacity for narrative is, as far as we know, unique to our species, because our species is, as far as we know, the only one that possesses self-consciousness. We see ourselves. Not only do we exist, but we think about our existence. We envision ourselves occupying a context — or, in storytelling terms, a setting: a place and a time. To see yourself as existing in a specific place and at a particular time is to suggest that you have existed and that you will exist in other places and at other times. You know you were born. You wonder what happens when you die.

But it's not just you that you wonder about. You take a walk and look at the stars, and because you know you are taking a walk and looking at the stars, you understand that you are joining a story already in progress. You ask yourself how it all got here. The answer you invent might involve light and dark, water and fire, semen and egg, gods or God, turtles, trees, trout. And when you have fashioned a sufficiently satisfying answer, you ask yourself, naturally, where it all — and you with it — will end. Bang? Whimper? Heaven? Nothing?

These questions might seem to lie outside the realm of physics, and before 1965 most scientists considered cosmology to be mostly that: metaphysics. Cosmology was where old astronomers went to die. It was more philosophy than physics, more speculation than investigation. The fourth member of the Princeton team — the one who didn't make the trip to Bell Labs — would have included himself in the category of cosmology skeptic.

Phillip James Edwin Peebles — Jim to everyone — was all angles.

Tall and trim, he explained himself to the world through his elbows and knees. He would throw his arms wide, as if to embrace every possibility, then wrap them around his legs, as if to consolidate energy and focus—mannerisms not inconsistent with a man of conflicting sensibilities, which was how Jim Peebles saw himself. Politically he called himself a "bleeding-heart liberal," yet scientifically he identified himself as "very conservative," even "reactionary." He had learned from his mentor, Bob Dicke, that a theory can be as speculative as you like, but if it doesn't lead to an experiment in the near future, why bother? On one occasion (before he knew better), Peebles had mentioned that he might try to reconcile the two great physics theories of the twentieth century, general relativity and quantum mechanics. "Go find your Nobel Prize," Dicke answered, "and then come back and do some real physics."

Cosmology, to Peebles, was not real physics. It was a reversion to how scientists did science in the two millennia before there were scientists and science as we know them. Ancient astronomers called their method "saving the appearances"; modern scientists might call it "doing the best you could under impossible circumstances." When Plato challenged his students, in the fourth century B.C., to describe the motions of the celestial bodies through geometry, he didn't expect the answers on paper to represent what was actually happening in the heavens. That knowledge was unknowable because it was unattainable; you couldn't go into the sky and examine it for yourself. What Plato wanted instead was an *approximation* of the knowledge. He wanted his students to try to find the maths to match not the facts but the appearances.

One student, Eudoxus, arrived at an answer that, in one form or another, would survive for two thousand years. For mathematical purposes he imagined the heavens as a series of nesting, concentric, transparent spheres. Some of these spheres carried the heavenly bodies. Others interacted with those spheres to retard or accelerate their motions, in order to account for the appearance that the heavenly bodies all slow down or speed up throughout their orbits. Eudoxus assigned the Sun and the Moon three spheres each. To each of the five planets (Mercury, Venus, Mars, Saturn, Jupiter) he assigned an extra sphere to accommodate the appearance that they sometimes briefly reverse their motions against the backdrop of stars, moving west to east from night to night rather than east to west.* And then he added a sphere for the realm of the stars. In the end his system consisted of twenty-seven spheres.

Another student of Plato's, Aristotle, amended this system. He assumed the spheres were not just mathematical constructs but physical realities; to accommodate the mechanics of an interlocking system, he added counterturning spheres. His total: fifty-six. Around A.D. 150, Ptolemy of Alexandria assumed the task of compiling the existing astronomical wisdom and simplifying it, and he succeeded: His night sky was overrun with only forty spheres. The maths still didn't match the appearances exactly, but close enough was good enough—as good as it was ever going to get.

Today, the 1543 publication of *De revolutionibus orbium coelestium* ("On the Revolutions of the Heavenly Spheres"), by the Polish astronomer Nicolaus Copernicus, is synonymous with the invention of a new universe: the Copernican Revolution. It has become a symbol of defiance against the Church's teachings. But it was the Church itself that had invited Copernicus to come up with a new maths for the motions in the heavens, and it had done so for a sensible reason: The appearances once again needed saving.

Over the centuries the slight inconsistencies in the Ptolemaic version—the areas where the maths departed from the motions—had led to a gradual drift in the calendar, until seasons diverged from their traditional dates by weeks. Copernicus's work allowed the Church to reform the calendar in 1582, incorporating his maths while dispensing with the notion of a Sun-centred universe. Like the ancients, Copernicus wasn't proposing a new universe, either physically or philosophically. Instead, he was formulating a new way to "save the appearances" of the existing universe. The true motions of that universe, however, were out of reach, had always been out of reach, and would always be out of reach.

And then, they weren't. In 1609, the Italian mathematician Galileo Galilei found new information about the universe at his finger-

* Today we would say that retrograde motion is the result of Earth overtaking another planet, or vice versa, in their orbits around the Sun.

tips—literally, thanks to the invention of a primitive telescope. *Look,* he said, leading the elders of Venice up the steps of the Campanile in the Piazza San Marco in August 1609 to demonstrate the benefit of fitting a tube with lenses: seeing farther. *Look,* he said barely six months later, in his pamphlet *Sidereus Nuncius* ("Starry Messenger"), heralding a new lesson: Seeing farther means seeing not just more of the same—a fleet of rival merchants or the sails of an enemy navy—but seeing more, full stop. That autumn, Galileo had trained his tube of longseeing on the night sky and had begun a lengthy program of discovering celestial objects that no other person had ever seen: mountains on the Moon, hundreds of stars, spots on the Sun, satellites of Jupiter, the phases of Venus. The invention of the telescope—the first instrument in history to extend one of the human senses—changed not only how far we could see into space, or how well. It changed our knowledge of what was out there. It changed the appearances.

Here was evidence that corroborated the central tenets of Copernicus's maths—that Earth was a planet, and that it and all the other planets orbited the Sun. But just as important, here was *evidence*—the tool of the scientific method. Seeing farther didn't have to mean seeing more. The night sky might not have held more objects than met the naked eye. And we still couldn't go into the sky and see for ourselves how its motions worked. But we could examine the heavens closely enough to find not only the appearances but the facts.* And facts needed not saving but explaining.

In 1687 the English mathematician Isaac Newton provided two of those explanations in *Philosophiæ Naturalis Principia Mathematica* ("Mathematical Principles of Natural Philosophy"). He reasoned that if Earth is a planet, then the formulae that apply in the terrestrial realm must apply in the celestial as well. Building on the mathematical work of Johannes Kepler and the observations of Galileo and his successors in astronomy, he concluded that the motions of the heavens require not dozens of spheres but a single law: gravitation. In 1705 his friend and sponsor Edmond Halley applied Newton's law to past observations of comets that had appeared in 1531, 1607, and 1682 to

* The distinction that eventually got Galileo into trouble with the Church.

make the claim that they were one comet and that it* would return in 1758, long after his own death. It did. No longer would the maths have to accommodate the motions of the heavens. Now the heavens had to accommodate the maths. Take Newton's law of universal gravitation, apply it to the increasingly precise observations you could make through a telescope, and you had a universe that was orderly and predictable and, on the whole, unchanging—a cosmos that ran, as the most common metaphor went, like clockwork.

In the more than three and a half centuries between Galileo's climb up the Campanile and the phone call from Crawford Hill, the catalogue of the universe's contents seemed to grow with every improvement of the telescope: more moons around planets; more planets around the Sun; more stars. By the early twentieth century, astronomers had determined that all the stars we see at night, whether with our naked eyes or through telescopes, are part of one vast collection of stars, numbering in the tens of billions, that we long ago named the Milky Way because it seems to spill across the night sky. Did other vast collections of stars, each numbering in the tens of billions, exist beyond the Milky Way? A simple extrapolation from the earlier pattern of discovery raised the possibility. And astronomers even had a candidate, a class of celestial objects that might qualify as "island universes" all their own.

In 1781, the French astronomer Charles Messier had published a catalogue of 103 celestial smudges—blurry objects that he feared would distract astronomers looking for comets. Astronomers could see that several of those 103 smudges were bunches of stars. As for the others, they remained mysteries, even as the quality of telescopes improved. Were these nebulous objects clouds of gas in the process of coalescing into yet more stars within our system? Or were the nebulae vast collections of tens of billions of stars separate from but equal in magnitude to our own vast collection? The astronomy community split on the question, and in 1920 two prominent astronomers conducted a so-called Great Debate at the National Museum of Natural History in Washington, D.C., to present the pros and cons of each argument.

* Yes, the eponymous Halley's Comet.

Three years later, the American astronomer Edwin Hubble did what debate alone couldn't do: resolve the question through empirical evidence. On 4 October 1923, using the largest telescope in the world—the new 1.5-metre* on Mount Wilson, in the hills outside Pasadena—he took a photograph of the Great Andromeda Nebula, or M31 in the Messier catalogue. He thought he noted a "nova," or new star, so he returned to M31 the following night and took another photograph of the same spiral arm. When he got back to his office, he began comparing the new plates with other photographs of the nebula on a number of different dates and found that the nova was actually a variable, a kind of star that, as its name suggests, varies: It pulsates, brightening and dimming with regularity. More important, it was a Cepheid variable, the kind that brightens and dims at regular time intervals. That pattern, Hubble knew, could resolve the debate.

In 1908, the Harvard astronomer Henrietta Swan Leavitt had discovered a proportional relationship between the pulsation period of a Cepheid variable and its absolute brightness: the longer the period, the brighter the variable. Astronomers could then take that measure of luminosity and match it with another quantifiable relationship, the one between luminosity and distance: A source of light that's twice as distant as another source of light with the same luminosity appears to be one-fourth as bright; a source of light three times as distant appears to be one-ninth as bright; a source of light four times as distant would be one-sixteenth as bright; and so on. If you know how often a variable pulsates, then you know how bright it is relative to other variables; if you know how bright it is relative to other variables, then you know how distant it is relative to other variables. When Hubble compared the pulsation period of the Cepheid variable he'd found in M31 with the pulsation periods of other Cepheid variables, he concluded that the variable was at sufficient distance that it (and therefore its host nebula, M31) lay beyond the "island universe"—or, as we would now have to think of it, *our* island universe.

Hubble went back to H335H, the photographic plate he made on 5 October, and in posterity-radiant red he marked the variable star with an arrow, along with a celebratory "VAR!" He declared M31 an

* The diameter of the light-collecting surface.

island universe all its own, and in so doing, he added to the cosmic canon one more *more:* galaxies.

Newton's clockwork universe began to come apart in 1929. After his "VAR!" breakthrough, Hubble had continued to investigate "island universes," especially some inexplicable measurements of them that astronomers had been making for more than a decade. In 1912 the American Vesto Slipher began examining the nebulae with a spectrograph, an instrument that registers the wavelengths from a source of light. Much like the sound waves of a train whistle as the train approaches or departs from a station, light waves are compressed or stretched—they bunch up or elongate—depending on whether the source of the light is moving toward you or away from you. The *speed* of the light waves doesn't change; it remains 186,282 miles (or 299,792 kilometres) per second. What changes is the *length* of the waves. And because the length of the light waves determines the colours that our eyes perceive, the colour of the source of light also seems to change. If the source of light is moving towards you, the waves bunch up, and the spectrometer will show a shift towards the blue end of the spectrum. If the source of light is moving away from you, the waves relax, and the spectrometer will show a shift towards the red end of the spectrum. And as the velocity of the source of light as it moves towards you or away from you increases, so does the blueshift or redshift—the greater the velocity, the greater the shift. Slipher and other astronomers had shown that some of the nebulae were registering significant redshifts, suggesting that they were moving away from us at great velocities. Now that Hubble knew these nebulae were galaxies, he wondered what these motions might mean. He found out when he compared the velocities of eighteen of these nebulae with their distances: The two measurements seemed to be directly proportional to each other—the farther the galaxy was, the faster it appeared to be receding. In other words, the universe might seem to be expanding.

Suddenly the universe had a story to tell. Instead of a still life, it was a movie. And like any narrative, the story of the universe now had not only a middle—the present, swarming with galaxies fleeing one another—but the suggestion of a beginning.

Precisely—*precisely*—at this point, at least from the perspective of a philosophically cautious sort like Jim Peebles, cosmology departed from science, passing from maths to myth. You couldn't know how the universe began because the evidence was out of reach, just as it had been for Aristotle, Ptolemy, and Copernicus. They couldn't go across space to retrieve it; you couldn't go into the past. All you could do was observe the present phenomena — these redshifted galaxies — and try to find the maths to accommodate their motions. All you could do was try to save the appearances, if that was your idea of science.

Hubble himself, as an observer, hoarding evidence and leaving the theorizing to the theorists, preferred to remain agnostic as to whether the universe really was expanding or whether another interpretation might explain the apparent correlation. But some theorists couldn't resist the challenge of rewinding the film. The Belgian priest Georges Lemaître, a physicist and astronomer, imagined the expansion unreeling in reverse, the size of the universe shrinking, smaller and smaller, the galaxies rushing back together, faster and faster, until the infalling matter would reach a state that he called the "primeval atom" and that other astronomers would come to call a "singularity": an abyss of infinite density and incalculable mass and energy.

But words such as "infinite" and "incalculable" aren't of much use to mathematicians, physicists, or other scientists. "The unrestricted repeatability of all experiments is the fundamental axiom of physical science," Hermann Bondi and Thomas Gold, two Austrian expatriates living in Britain, wrote in the first line of a paper they submitted in July 1948 that outlined an alternative to Lemaître's theory. The following month, their friend Fred Hoyle, a British astronomer, submitted his own variation on this theme. Rather than a big bang — the term Hoyle applied, during a BBC radio broadcast in March 1949, to the idea of a universe expanding* from, as he wrote in his paper, "causes unknown to science" — they postulated a steady state. Through "continuous creation of matter," Hoyle wrote, "it might be possible to obtain an expanding universe in which the proper density of matter remained constant." Over the course of cosmic history, the

* Technically the term applies to the expansion — to everything that has happened after the singularity — though through common usage it has also come to mean the singularity itself.

creation of even infinitesimal amounts of matter could become cumulatively significant. Such a universe wouldn't have a beginning or an end; it would just *be.*

For many astronomers, however, "continuous creation" was no more appealing than a "singularity." Both the Big Bang and Steady State theories seemed to require a leap of faith, and faith not being part of the scientific method, there they let the matter rest.

But what if there *was* evidence for one theory or the other?

Bob Dicke asked Jim Peebles this question one sweltering evening in 1964. Peebles had arrived at Princeton as a graduate student six years earlier. At the University of Manitoba he had been the top student in physics, winning academic honour after honour. At Princeton he was shocked at how much physics he didn't know. He spent his first year trying to catch up, and then one day some friends invited him to a get-together that Dicke ran most Friday evenings in the attic of Palmer Physical Laboratory. The Gravity Group was an informal gathering of a dozen or so undergraduates, graduates, postdocs, and senior faculty—"Dicke birds," they called themselves. Peebles went, and then he went back. He began to understand that here, in a sometimes-stifling setting at an inconvenient hour, he could get an education: eating pizza, drinking beer, and trying to figure out how to rehabilitate general relativity.

General relativity had been around for nearly half a century; Einstein had arrived at the equations in late 1915. Whereas Newton imagined gravity as a *force* that acts *across* space, Einstein's equations cast gravity as a *property* that belongs *to* space. In Newton's physics, space was passive, a vessel for a mysterious force between masses. In Einstein's physics, space was active, collaborating with matter to produce what we perceive as gravity's effects. The Princeton physicist John Archibald Wheeler offered possibly the pithiest description of this co-dependence: "Matter tells space how to curve. Space tells matter how to move." Einstein in effect reinvented physics. Yet by 1940 Dicke could ask a professor of his at the University of Rochester why the graduate physics curriculum didn't include general relativity, and the answer was that the two had nothing to do with each other.

Einstein might have agreed. A sound theory needs to make at least one specific prediction. General relativity made two. One involved an infamous problem of Einstein's era. The orbit of Mercury seemed to be slightly wrong, at least according to Newton's laws. The observable differences between Newton's and Einstein's versions of gravity were negligible — except in circumstances involving the most extreme cases, such as a tiny planet travelling close to a gargantuan star. Newton's equations predicted one path for Mercury's orbit. Observations of Mercury revealed another path. And Einstein's equations accounted precisely for the difference.

Another prediction involved the effect of gravity on light. A total eclipse of the Sun would allow astronomers to compare the apparent position of stars near the rim of the darkened Sun with their position if the Sun weren't there. According to general relativity, the background starlight should appear to "bend" by a certain amount as it skirted the great gravitational grip of the Sun. (Actually, in Einstein's theory it's space itself that bends, and light just goes along for the ride.) In 1919 the British astronomer Arthur Eddington organized two expeditions to observe the position of the stars during an eclipse on 29 May, one expedition to Principe, an island off the west coast of Africa, the other to Sobral, a city in northeastern Brazil. The announcement in November 1919 that the results of the experiments seemed to validate the theory made both Einstein and general relativity international sensations.*

Yet Einstein himself downplayed the theory's power to predict "tiny observable effects" — its influence on physics. Instead, he preferred to emphasize "the simplicity of its foundation and its consistency" — its mathematical beauty. Mathematicians tended to agree, as did physicists such as Dicke's professor at the University of Rochester. General relativity's known effects in the universe — an anomaly in the orbit of a planet, the deflection of starlight — were obscure in the extreme; its unknown effects on the history of the universe — cosmology — were speculative in the extreme. Even so, Einstein also acknowledged that if the theory made a prediction that observations

* Einstein thought his theory made a third prediction, involving the redshifting or blueshifting of light by gravity, but it turned out not to be specific to general relativity.

contradicted, then, as would be the case with any theory under the standards of the scientific method, science should amend or abandon it.

By the time Dicke joined the Princeton faculty in the 1940s, after the war, Einstein was as much a spectral presence in life as his theory was in experimental physics. Sometimes a seemingly homeless man would shuffle into a faculty party, and the younger folks in the crowd would need a moment to recognize the shock of hair and the basset-hound eyes. During the 1954–55 academic year, Dicke took a sabbatical leave at Harvard, and he found himself returning to thoughts of general relativity. As a scientist who was equally at ease designing equipment and constructing theories, Dicke realized he could do what previous generations couldn't have done with their existing technology. When he returned to Princeton, he resolved to put Einstein to the test.

His experiments over the coming years would involve placing occulting disks in front of the Sun to determine its precise shape, which affects its gravitational influence on the objects in the solar system, including Mercury; bouncing lasers off the Moon and using the round-trip time to measure its distance from Earth, which would indicate if its orbit was varying from Einstein's maths in the same way that Mercury's orbit varied from Newton's maths; and using the chemical composition of stars to trace their age and evolution, which in turn would be important for tracing the age and evolution of the universe, which in turn would involve an attempt to detect the relic radiation from the primeval atom, cosmic fireball, Big Bang, or whatever you wanted to call it. Dicke wondered if a theory of the universe could avoid not only a Big Bang singularity but the Steady State's spontaneous creation of matter, and he proposed a compromise of sorts: an oscillating universe.

Such a universe would bounce from expansion to contraction to expansion throughout eternity, without ever reaching absolute collapse or, between collapses, eternal diffusion. During the expansion phase of such a universe, galaxies would exhibit redshifts consistent with what astronomers were already observing. Eventually the expansion would slow down under the influence of gravity, then reverse itself. During the contraction phase, galaxies would exhibit blueshifts as they gravitated back together. Eventually the contraction would reach a state of such compression that it would explode outwards again, before the laws of physics broke down. Dicke's oscillating universe would therefore neither emerge from nor return to the dreaded singularity, though the earliest period of its current expansion would resemble a Big Bang. During one particularly muggy meeting of the Gravity Group, Dicke ended a discussion of that theory by turning to two of his Dicke birds, Peter Roll and David Todd Wilkinson, and saying, "Why don't you look into making a measurement?" They could build a radio antenna to detect the radiation from the most recent Big Bang. Then he turned to a twenty-nine-year-old postdoc and said, "Why don't you go think about the theoretical consequences?"

Jim Peebles had already forced himself to learn cosmology. As a graduate student at Princeton he had been required to pass the Physics Department's general examinations, and when he looked at previous years' exams, he saw that they reliably included questions on general relativity and cosmology. So he studied the standard texts of the day, *Classical Theory of Fields,* by Lev Landau and Evgeny Lifschitz, from 1951, and *Relativity, Thermodynamics, and Cosmology,* by Richard C. Tolman, from 1934. Both books came with a whiff of formaldehyde; they presented cosmology in the embalmed terms of long-settled truths. The more Peebles educated himself about cosmology, the less he trusted it. General relativity itself excited him; he was a loyal and enthusiastic member of the Gravity Group. What appalled him were the assumptions that theorists had forcibly yoked to general relativity in order to create their cosmologies.

The trouble, Peebles saw, started with Einstein. In 1917, two years after arriving at the theory of general relativity, Einstein published a paper exploring its "cosmological considerations." What might general relativity say about the shape of the universe? In order to simplify the maths, Einstein had made an assumption: The distribution of matter in the universe was homogeneous—that is, uniform on a large scale. It would look the same no matter where you were in it. In calculating the implications of Einstein's theory, Georges Lemaître and, independently, the Russian mathematician Aleksandr Friedman had adopted the same assumption and added one more, that the universe is isotropic—uniform in every direction. It would look the same no matter which way you looked. Then the Steady State theory went the Big Bang one assumption better: The universe is homogeneous and isotropic not only throughout space but over time. It would look the same in every direction no matter where you were in it and no matter when.

Peebles tried to be fair; he attended a lecture on the Steady State theory. But he came away thinking, "They just made this up!" To Peebles, a homogeneous universe, whether in space or time or both, was not a serious model. Tolman's book came right out and said as much: Theorists assumed homogeneity "primarily in order to secure a definite and relatively simple mathematical model, rather than to secure a correspondence to known reality." This approach reminded Peebles of those oversimplified problems on exams: *Calculate the acceleration of a frictionless elephant on an inclined plane.*

"This is silly," Peebles thought. Why, he asked himself, would anyone imagine the universe to be, of all the things that a universe could be, *simple?* Yes, scientists preferred to follow the principle of Ockham's razor, dating back to the fourteenth-century Franciscan friar William of Ockham: Try the simplest assumptions first and add complications only as necessary. So Einstein's invocation of a homogeneous universe had a certain logic to it, a legacy behind it—but not enough to be the basis of a science that made predictions that led to observations.

Yet when Dicke approached him about figuring out the temperature of the most recent Big Bang in an oscillating universe, Peebles immediately accepted the challenge. First, the request came from Bob Dicke, and you had to trust his hunches. Besides, Peebles shared not only his mentor's enthusiasm for exploring general relativity but Dicke's reservations about cosmology. Only a year earlier, in 1963, in an article on cosmology and relativity for the *American Journal of Physics,* Dicke had written: "Having its roots in philosophic speculations, cosmology evolved gradually into a physical science, but a science with so little observational basis that philosophical considerations still play a crucial if not dominant role."

What appealed to Peebles was the chance to shore up that "observational basis"—the experimental implications. It was the possibility that his calculations might lead to an actual measurement, one that

Roll and Wilkinson would make, using the radio antenna that Dicke had assigned them to build. They would be doing cosmology the scientific way: The appearances were going to have to accommodate Jim Peebles's maths.

The first hint that radio waves might offer a new way of seeing the universe dated to the 1930s—again, through an accidental detection at Bell Labs. In 1932 an engineer who had been trying to rid transatlantic radiotelephone transmissions of mystery static figured out that the noise was coming from the stars of the Milky Way. The news made the front page of the *New York Times* but then receded into obscurity. Even astronomers regarded the discovery as a novelty. Not until after World War II did the use of radio waves to study astronomy become widespread.

Radio astronomy turned out to be part of a larger dawning of awareness among astronomers that the range of the electromagnetic spectrum beyond the narrow optical band might contain useful information. The wavelengths to which human eyes have evolved to be sensitive range from 1/700,000th of a centimetre (red) to 1/400,000th of a centimetre (violet). To either side of that narrow window of sight, the lengths of electromagnetic waves increase and decrease by a factor of about one quadrillion, or 1,000,000,000,000,000. The Princeton experiment would concentrate on some of the longest waves because they would have the lowest energy—the kind that radiation that had been cooling from very nearly the beginning of time would have reached by now.

Peebles began by using the present constitution of the universe to work backwards towards the primordial conditions. The present universe is about three-quarters hydrogen, the lightest element; its atomic number is 1, meaning that it has one proton. In order for such an abundance of hydrogen to have survived to the present day, the initial conditions must have contained an intense background of radiation, because only an extraordinarily hot environment could have fried atomic nuclei fast enough to keep all those single protons from fusing with other subatomic particles to form helium and heavier elements. As the universe expanded—as its volume grew—its temperature fell. Extrapolate from the current percentage of hydrogen how intense the initial radiation must have been, calculate how much the volume of the universe has expanded since then, and you have the temperature to which the initial radiation would by now have cooled.

A radio antenna, however, doesn't measure temperature, at least not directly. The temperature of an object determines the motions of its electrons—the higher the temperature, the greater the motions. The motions of the electrons in turn are what produce radio noise—the greater the motions, the more intense the noise. The intensity of the noise therefore tells you how much the electrons are moving, which tells you the temperature of the object—or what engineers call the "equivalent temperature" of the radio noise. In a box with opaque walls, the only source of radio noise will be the motions of the electrons in the walls. If you place a radio receiver in a box that happens to be the universe, then the intensity of the static will tell you the equivalent temperature of the walls of the universe: the relic radiation.

In 1964 Peebles got to work predicting the current temperature of the relic radiation—the equivalent temperature of the static that an antenna would need to detect. Meanwhile, his colleagues Roll and Wilkinson began work on the antenna—technically, a Dicke radiometer, invented by Dicke to refine radar sensitivity during the war, while he was working at the Radiation Laboratory at the Massachusetts Institute of Technology. In early 1965, Peebles received an invitation from Johns Hopkins University's Applied Physical Laboratory to give a talk, and he asked Wilkinson if he could mention the radiometer in public.

"No problem," Wilkinson said. "No one could catch up with us now."

What happened next happened fast. Peebles delivered his talk on 19 Februar y. In the audience was a good friend of Peebles's from his graduate studies (and a former Dicke bird), Kenneth Turner, a radio astronomer at the Carnegie Institution's Department of Terrestrial Magnetism (DTM), in Washington, D.C. The experiment made an impression on Turner, and a day or two later he mentioned the colloquium to another radio astronomer at DTM, Bernard Burke. Another day or two later, during a communal lunch, Burke got a phone call from a Bell Labs radio astronomer he'd met in December on a plane ride to an American Astronomical Society meeting in Montreal. Burke went into the kitchen's anteroom to take the call. After a brief discussion, Burke made small talk. "How is that crazy experiment of yours coming?" he said.

On the flight to Montreal, Arno Penzias had described for Burke the work he and Bob Wilson were doing on Crawford Hill. He had told Burke that they hoped to study the radio waves from the stars not in the big bulge at the centre of the Milky Way, where most astronomers had been looking, but in the other direction, at the fringe of the Milky Way halo. But now, he said to Burke on the phone, they'd run into a problem even before they could begin their observations.

"We have something we don't understand," Penzias said. He explained that he and Wilson couldn't get rid of an excess noise corresponding to a temperature near, but not quite, absolute zero. When Penzias had finished describing their efforts and their frustration, Burke said, "You should probably call Bob Dicke at Princeton."

The Big Bang was a creation myth, but by 1965 it was a creation myth with a difference: It came with a prediction. By the time Penzias placed his phone call to Dicke, Peebles had arrived at a temperature of approximately 10° Celsius above absolute zero, which is more commonly referred to as 10 Kelvin.* Penzias and Wilson had found a measurement of 3.5 K (plus or minus 1 K) in their antenna. Because Peebles's calculations were rudimentary and Penzias and Wilson's detection was serendipitous, the approximation of theory and observation was hardly definitive. Yet it was also too close to dismiss as coincidence.

At the very least it was worth recording for posterity. After the Crawford Hill meeting and a reciprocal meeting at Princeton, the two sets of collaborators agreed that they would each write a paper, to appear side by side in the *Astrophysical Journal*. The Princeton foursome would go first, discussing the possible cosmological implications of the detection. Then the Bell Labs duo would confine their discussion to the detection itself, so as not to align their measurement

---

* Absolute zero, in principle the coldest temperature possible, is –273.15° Celsius. By convention, scientists designate absolute zero as 0 Kelvin and count upwards in increments of degrees Celsius. So 10°C above absolute zero is 10 K.

too closely with a wild interpretation that, as Wilson said, it "might outlive."

On 21 May 1965, even before their papers appeared, the *New York Times* broke the story: "Signals Imply a 'Big Bang' Universe." (The reporter had been in contact with the *Astrophysical Journal* about another upcoming paper when he heard about these two papers.) The prominence of the coverage—placement on the front page; an accompanying photograph of the Bell Labs telescope—impressed some of the scientists in the two collaborations with the possible impact of their (possible) discovery. Peebles, though, didn't need the news media to tell him they were onto something big. All he had to do was look at Dicke. Dicke could be humourous and lighthearted, but not about physics. In recent weeks, though, he had clearly been enjoying himself in a different way. After talking to Dicke, one longtime Princeton astronomer reported back to his peers that Bob Dicke was "bubbling with excitement."

A subsequent search of the literature turned up other predictions and at least one previous detection. In 1948, the physicist George Gamow had written a *Nature* paper that predicted the existence of "the most ancient archeological document pertaining to the history of the universe." He was wrong on the details but right on the general principle: The early universe had to be extremely hot to avoid combining all the hydrogen into heavier elements. That same year, the physicists (and sometime collaborators of Gamow's) Ralph Alpher and Robert Herman published their calculation that "the temperature in the universe" should now be around 5 K, but astronomers at the time assured them that such a detection would be impossible with current technology. (In retrospect Wilson felt that they probably could have performed it with World War II–era technology, as long as they had properly connected the antenna to the cold load.) In a 1961 article in the *Bell System Technical Journal*, a Crawford Hill engineer wrote that the Echo antenna was picking up an excess of 3 K; but that reading fell within the margin of error, and the discrepancy wasn't going to make a difference for his purposes anyway, so he ignored it. In 1964, Steady State champion Hoyle, working with fellow British astronomer Roger J. Tayler, investigated the oscillating-universe scenario and performed calculations similar to those of Alpher and Herman. Also in 1964, even as Penzias and Wilson were directing their antenna to every point on the horizon in a futile effort to find the source of their excess noise, two Russian scientists published a paper pointing out that a detection of the cosmic background radiation was currently possible—and that the ideal instrument was a certain horn antenna on a hilltop in Holmdel Township, New Jersey.

Jim Peebles had a high metabolism; he could eat whatever he wanted and not worry about gaining weight. This inherent restlessness extended to his intellectual life. He loved identifying the next big problem, solving it, seeing where it led, identifying *that* big problem, solving it, seeing where it led: a bend-in-the-knees, wind-in-the-face rush into the future. (He was an expert downhill skier.) Even the description of his intellectual restlessness that he once gave to a journalist was restless: "a random walk, no, an undirected walk, or rather a locally directed walk: as you take each step you decide where the next one is going to go." The library part of the scholarly process, however, the burrowing into the stacks, the boning up on the literature—maybe it didn't bore him, exactly, but it didn't engage him either. In any case, he hadn't done his homework.

His initial paper on the temperature of the universe—Dicke had forwarded a preprint to Penzias after the phone call about the Bell Labs detection—had repeatedly bounced back from the *Physical Review* referee because it was duplicating earlier calculations by Alpher, Herman, Gamow, and others. Peebles finally withdrew the paper in June 1965. He managed to rectify some of those oversights in the paper on the cosmic microwave background he wrote with Dicke, Roll, and Wilkinson. Even that paper, however, referred only to Gamow's work on the primordial creation of elements, not to his work predicting the temperature of the cosmic background. Gamow sent an angry note to Penzias, listing citations of his early work and concluding, "Thus, you see, the world did not start with almighty Dicke."

Still, the obscurity in which these documents languished was a reflection of the indifference many scientists felt toward cosmology and general relativity. No longer. By December 1965, Roll and Wilkinson had mounted their antenna on the roof of Guyot Hall and gotten the same reading as Penzias and Wilson. Within months two more experiments (one by Penzias and Wilson) had found what a sound scientific prediction demands: a duplication of results—in this case, a detection of what was already being called "the 3-K radiation."

You could feel the shift, if you were an astronomer or physicist. Both the Steady State and Big Bang interpretations had relied not just on maths and observation but on speculation. They were modern counterparts to Copernicus's attempt to save the appearances; they were theories in need of evidence. And just as Galileo, with the aid of the telescope, had detected the celestial phenomena that decided between an Earth-centred and a Sun-centred cosmos, forcing us to reconceive the universe, so radio astronomers, with the aid of a new kind of telescope, were now detecting the evidence that decided between the Steady State and Big Bang cosmologies, necessitating a further reconception of the universe.

Seeing beyond the optical part of the electromagnetic spectrum didn't have to mean seeing more. The sky might not have harboured more information than meets the eye, even one aided by an optical telescope. The introduction of radio astronomy could have left the Newtonian conception of the universe intact. But seeing beyond the optical did mean seeing more phenomena and having to accommodate new kinds of information. This new universe would still run like clockwork; the laws that had arisen through Galileo's observations and Newton's computations would still presumably apply. But now, so would Hubble's and Einstein's, and in their universe the motions of the heavens weren't cyclical so much as linear; their cosmos corresponded not so much to a pocket watch, its hands and gears grinding and turning but always returning to the same positions, as to a calendar, its fanning pages preserving the past, recording the present, and promising the future.

Maybe, Peebles thought, making theories of the universe wasn't so silly after all. Not that the always-cautious Peebles now embraced the Big Bang theory. But the uniformity of the microwave background that he had predicted and that Penzias and Wilson had detected would certainly correspond to a universe that looked the same on the largest scales no matter where you were in it. Einstein had posited an elephant on an incline, and that's what the universe turned out to be: homogeneous.

"Which is an amazing thing," Peebles thought. "But there it is: The universe is simple."

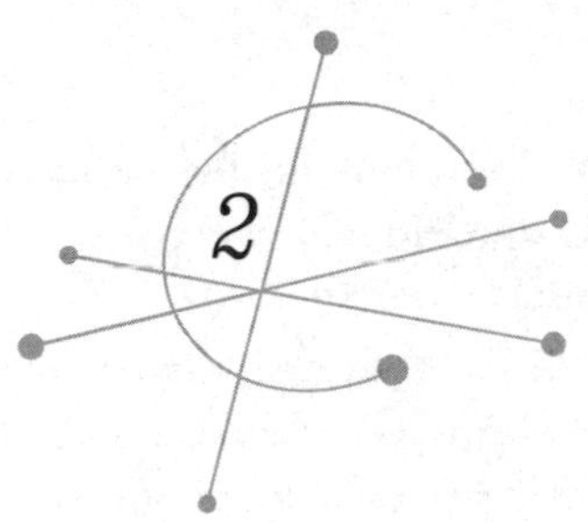

# What's Out There

WHAT THE UNIVERSE could be, or should be, didn't much concern her. She wasn't a theorist. She was an astronomer — an observer. The universe was what it was. And what it was, everywhere you looked, was in motion.

Well, she wasn't an astronomer *yet*. She'd never actually observed, except as a child, using a telescope that her father had helped her build out of a lens she'd ordered through the post and a cardboard linoleum tube she'd gotten free from a shop in downtown Washington, D.C. And that telescope didn't even work properly; she couldn't take pictures of the stars with it, because it couldn't track their motions — or, more accurately, their apparent motions, since it's the turning of the Earth that gives stars the illusion of arcing across the night sky.

She should have known that the camera wouldn't work. The motions of the stars were part of what got her interested in astronomy. Her second-floor bedroom window — the one right above her bed — faced north, and around age ten she noticed that the stars appeared to be slowly circling a point in the northern sky, and that over the seasons the stars themselves changed. Ever since, she found that she she would rather track the motions of the night sky than sleep. She memorized the paths of meteors, then registered them in a notebook in the morning. Later, in secondary school, whenever she had to write a research paper the topic she chose was invariably something to do with astronomy — reflecting telescopes (the kind with mirrors) or refracting telescopes (the kind with lenses). At a certain point in the evening her mother might call up the stairs, "Vera, don't spend the whole night with your head out the window!" But she did, and her parents didn't seem to mind, not really.

Hers was, in a way, a Newtonian view of the universe: matter in motion; predictable patterns; celestial objects (and the Earth was one, too, if you thought about it) that, for all their peregrinations, invariably wound up back where they started. But Vera Cooper was born in 1928, three years after Edwin Hubble announced that our Milky Way galaxy was hardly singular, and one year before he presented evidence that the galaxies seemed to be receding from one another—the farther apart, the faster. The only universe she'd known was full of galaxies, and those galaxies were in motion.

And so, as a graduate student at Cornell, when she had to think about a topic for her master's thesis, she tried to update the old clockwork view of the cosmos for the new expanding universe. She reasoned that since the Earth rotated on its axis, and the solar system rotated, and the galaxy rotated, then maybe the universe had an axis too. Maybe the whole universe rotated.

The premise seemed reasonable. Her husband, Robert Rubin, a PhD candidate in physics at Cornell, had shown her a brief, speculative article by George Gamow in the journal *Nature,* "Rotating Universe?" Then she heard that Kurt Gödel, at Princeton, was working on a theory of a rotating universe.

Her approach also seemed reasonable. She gathered data on the 108 galaxies for which astronomers had managed to measure a redshift. Then she separated out the motions that were due to the expansion of the universe—what astronomers call recessional motions. Did the motions that remained—the peculiar motions—exhibit a pattern? She plotted them on a sphere and thought they did. In December 1950, aged twenty-two, still half a year shy of getting her master's degree, Vera Cooper Rubin presented her thesis at an American Astronomical Society meeting in Haverford, Pennsylvania.

Rubin had never suffered from a lack of confidence. When an admissions officer at Swarthmore College told her that because astronomy was her profession of choice and painting was one of her favorite hobbies, she might want to consider a career as a painter of astronomical scenes, she laughed and applied to rival Vassar. When she got a scholarship to Vassar and a teacher told her, "As long as you stay away from science, you should do okay," she shrugged and pursued a degree in astronomy (with a heavy load of philosophy of science on the side). When a Cornell professor told her that because she had a one-month-old son he would have to take her place at the Haverford AAS meeting and present her paper in his own name, she said, "Oh, I can go," and, nursing newborn and all, she went.

The response from the AAS crowd when she concluded her presentation was nearly unanimous: The premise was odd, the data weak, the conclusion unconvincing. The criticism continued until the astronomer Martin Schwarzschild kindly signaled an end to the discussion by rising and saying, in a high-pitched voice, "This is a very interesting thing to have attempted." The moderator called a coffee break, and Rubin left the meeting.

She herself hadn't thought her paper was extraordinary; it was a master's thesis, after all. Still, she thought that as master's theses went, it was fine. She had taken a pile of numbers and handled them in the most careful fashion she knew, and she thought the result was worth reporting. She thought that she'd given a good talk, and that she'd given it as well as she could. She reminded herself that she had never been to an AAS meeting before, and that she hadn't even met many professional astronomers. Maybe this was just how astronomers behaved. She decided she would file these criticisms in the same category as the comments from the admissions officer and her teachers. The next day her hometown paper the *Washington Post* ran an article under the headline "Young Mother Figures Center of Creation by Star Motions." So she could console herself that real astronomers would at least know who she was (or, because of a typo, who "Vera Hubin" was).

Still, the experience did teach Rubin an important lesson: She was such a novice that she didn't know how far out of the mainstream her work was. She didn't know that Gamow was nearly alone among astronomers, and Gödel among theorists, in finding the question of a rotating universe worthy of serious consideration. Gamow had admitted, in the *Nature* paper, that the idea of a rotational universe was "at first sight fantastic"—which, at first sight, it was. But what if you didn't trust first sight? First sight—the evidence of the senses, unaided by technology—tells you that the Earth is stationary, that the Sun revolves around the Earth, that Jupiter is moonless and Saturn ringless and the stars motionless, and that the stars are as far as there is. The point Gamow was trying to make was that astronomers needed to go beyond first sight, because now they had a new scale of the universe to consider.

Saying that all the billions of stars we see are part of our galaxy and that billions of galaxies lie beyond our own doesn't do justice to the scale of the universe. Just as our eyes didn't need to evolve to see radio waves in order for us to survive, maybe our minds didn't need to evolve to understand the numbers that astronomers were now trying to incorporate into their thinking. Like cultures that count "One, two, three, more," we tend to regard the scale of the universe—to the extent that we regard it at all—as "Earth, planets, Sun, far."

Consider: How long would it take you to count to a million at the rate of one second per number? Eleven days—or, to be exact, 11 days, 13 hours, 46 minutes, and 40 seconds. How long would it take you to count to a billion at the same rate? A billion is a thousand million—that is, a million one thousand times over. So you would have to count for eleven thousand days. That is 31 years, 8½ months. To reach a trillion, you'd have to count to a billion a thousand times—31 years a thousand times, or 31,000 years. A light-year—the distance light travels in a year—is about six trillion miles. To count to six trillion, you would need six sets of 31,000 years, or 186,000 years.

Earlier generations of astronomers had to learn to adjust their thinking to accommodate successive discoveries about new scales of the universe: that the Sun is 93 million miles distant; that the nearest star after the Sun is 4.3 light-years, or 25 trillion miles, away (that's 186,000 years of counted seconds 4.3 times, or about 800,000 years); that our "island universe" consists of billions of stars at similar distances from one another; and that the diameter of this island universe, from one end of its spiral disk to the other, is about 100,000 light-years (186,000 years of seconds a hundred thousand times, or more than eighteen billion years of counting, a number you couldn't appreciate without first appreciating the meaning of "billion").

In this context, however unfathomable and even ludicrous, the term "billions of galaxies" at least begins to suggest the difference in scale between the island universe Hubble inherited and the universe he bequeathed to the next generation. His universe was saturated with galaxies as far as the "eye" could see—whether the "eye" was the one he used, the behemoth 2.5-metre Hooker telescope atop Mount Wilson, or its successor, the 5.1-metre Hale telescope atop Mount Palomar, which saw first light in 1949, promising astronomers access to galaxies at greater and greater redshifts. Who knew where this emerging reconception of the universe might lead? Astronomers of the mid-twentieth century who wanted to work in Hubble's universe would have to engage with its history and structure on the grandest scale imaginable. They'd have to do cosmology.

Not that Rubin thought of herself as a cosmologist. She didn't even think of herself as an astronomer, and not just because she'd never looked through a professional telescope. Six months after that AAS meeting, she had her master's degree and her husband had his PhD, and they had moved to the D.C. area to be near his job. Their son wasn't yet one, and they were planning to have another child, and even though her husband kept encouraging her to pursue her PhD, she felt that life was complicated enough at the moment ment. So it was her choice not to become an astronomer just yet, and to wonder every day whether she would ever become one. Even so, she felt that nothing had prepared her for this life: living in a suburb, staying home with her son while her husband went to work, crying whenever an issue of the *Astrophysical Journal*—she'd kept the subscription—came into the house.

Then one day the phone rang. It was George Gamow.

She was standing at the window in her flat. The phone rested on a table. The sofa was elsewhere. There was no place to sit. Did the cord stretch? No matter. It was the kind of conversation you wanted to have while standing. So she stood and stared out the window and listened as George Gamow asked Vera Rubin about her research.

Her husband shared an office with Ralph Alpher at Johns Hopkins University's Applied Physics Laboratory in Silver Spring, Maryland. Robert Herman had an office down the hall. Gamow did some consulting work for the laboratory, and Alpher and Herman often collaborated with him. From them Gamow had heard about her thesis. He told her that he wanted to know about her work on the rotation of the universe for a talk he was giving at the lab. (She wouldn't be able to attend: No wives allowed.)

Robert Rubin had taken the job at the Applied Physics Laboratory because the proximity to Washington might give his wife educational or professional options in astronomy—options she hadn't yet explored. After the phone call from Gamow, she started taking *ApJ* to the playground, and by February 1952, pregnant with her second child she was attending lectures at Georgetown University, the only school in the D.C. area to offer a doctorate in astronomy. There, by special arrangement with George Washington University, she would be working under the supervision of Professor Gamow.

That spring she met Gamow for the first time. He had suggested they meet in the library at the Carnegie Institution's Department of Terrestrial Magnetism, a modest campus on a hilly, wooded outskirt of Rock Creek Park in northwest Washington. DTM was an unassuming brick building in an unlikely setting. At the top of a hill at the end of a long, curving driveway in a residential neighborhood, it could have been a hospital or old peoples' home. Instead, beginning in the first decade of the century, it had been the headquarters of worldwide expeditions to chart the Earth's magnetic field; once that mission was completed, in 1929, DTM had adopted a looser interpretation of investigating the nature of our planet, and began research on nuclear physics and the geology of other planets.

Rubin had visited the sylvan setting on one earlier occasion—a lecture of some sort, probably. Now she found herself returning just about every month. The entrance to the library was to the right of the stairs on the second floor. To get from the door of the library to the reading room you had to squeeze through a narrow passage between two sets of bookshelves. Every time she visited she had to hesitate on the threshold, assessing one more challenge to becoming an astronomer. The passage was, perhaps, two feet wide. Pregnant with her second child, she was, perhaps, wider.

George Gamow turned out not to be the sort of person she might have hoped. When they didn't meet at the quiet, wood-paneled DTM library, they met at his home in Chevy Chase, Maryland. There he would invariably be shouting abuse at his wife in some distant part of the house. Where were his papers? What had she done with his papers? Why was she always going through his papers? Whether Gamow's wife was ever actually there, Rubin couldn't be sure. In the summer of 1953, Rubin and her husband paid their own way to an astronomy workshop in Michigan; Gamow was there too, and his behaviour embarrassed her. He dozed during talks, and when he woke up he asked questions that had already been answered. During afternoon discussions, just her and him and the great Mount Wilson astronomer Walter Baade, Gamow would down half a bottle of liquor. During his own lecture, he perspired alcohol.

Rubin was beginning to realize that there are two kinds of geniuses. There's the kind we would all be if we were extremely smart and knew what we were doing. And then there's the kind you can only watch, knowing that your mind could never work that way. That's the kind of genius Gamow was. He may have dozed during lectures and asked redundant questions, but he also answered questions that nobody else could answer. Whatever Gamow's personal failings, when he spoke, you listened.

"Is there a scale length in the distribution of galaxies?" he said to her at one of their first meetings. He was suggesting that she think not just about the overall motions of galaxies, as she had in her master's thesis, but instead about the overall result of those motions: the *arrangement* of galaxies.

Was the distribution of galaxies throughout the universe random and uniform, as most astronomers assumed? Hubble himself had thought so. "On a large scale the distribution is approximately uniform," he had written in his highly influential 1936 book, *The Realm of the Nebulae*. "Everywhere and in all directions, the observable region is much the same." In a sense, he was simply reiterating the two assumptions of modern cosmology, homogeneity and isotropy, in layman's terms. But the way he was framing the issue was also reminiscent of the premodern island-universe thinking—emphasis on "island." In Hubble's view, and therefore the view of a generation of astronomers, the galaxy clusters that astronomers had observed would be accidents of nature, or perhaps a sort of cosmic optical illusion arising from multiple galaxies falling along our line of sight.

But Gamow was thinking on a different scale. Maybe the peculiar motions of galaxies—the motions that were separate from a straightforward outwards expansion—weren't random, as most astronomers assumed. Maybe the gravitational interactions among galaxies, even across previously unthinkable distances, were sometimes strong enough to counteract the expansion on a local level. Maybe no—or at least not every—galaxy is an island.

The premise seemed respectable to Rubin, and not just because the visionary George Gamow was suggesting it. Shortly after that first phone call from Gamow, she had received a letter from Gérard de Vaucouleurs—a French astronomer then working in Australia—and then she heard from him again, and then again. She found the correspondence relentless; she always seemed to owe him a letter. But she couldn't complain. As was the case with Gamow, de Vaucouleurs wanted to discuss her master's thesis. He wrote to her that he had noticed a pattern among the galaxies similar to the one she had possibly detected, and in February 1953, midway through her PhD work, her patience with the persistent de Vaucouleurs paid off. He began an article in the *Astronomical Journal* with a citation from her work: "From an analysis of the radial velocities of about a hundred galaxies within 4 megaparsecs Mrs. V. Cooper Rubin recently found evidence for a differential rotation of the inner metagalaxy." To de Vaucouleurs, however, her evidence seemed to suggest not the rotation of the universe but the motion of a cluster of galaxy clusters—a supercluster. Even so, his argument was a variation on the theme Gamow was now asking Rubin to consider: Did galaxies cluster, and if so, why?

Once again she marshalled the data that was already out there, available to anyone, this time galaxy counts from Harvard. And once again she applied a conceptually straightforward analysis, this time plotting the galaxies in three dimensions by comparing locations in the sky with distances suggested by their redshifts.

She learned to balance becoming an astronomer and being a mother, sometimes literally: a thick German textbook in one hand, the handle of a baby pram in the other. Two or three evenings a week she attended lectures in the observatory at Georgetown. She worked on her thesis at night, after the children were in bed. She finished her studies in two years, and her thesis, "Fluctuations in the Space Distribution of the Galaxies," appeared in the 15 July 1954, issue of *Proceedings of the National Academy of Sciences.* Her conclusion: Galaxies don't just bump and clump arbitrarily; they gather for a reason, and that reason is gravity.

This time she didn't receive a drubbing, as she had at Haverford. The reaction was worse: silence.

During an AAS meeting in Tucson in 1963, Vera Rubin went on a tour of the Kitt Peak National Observatory, in the desert mountains fifty-five miles southwest of the city. Rubin was by then the mother of four and an assistant professor in astronomy at Georgetown, but she was still not a practicing astronomer. "Galaxies may be pretty remarkable," she liked to explain, "but to watch a child from zero to two is just incredible." Her youngest, however, was now three.

She secured time at Kitt Peak later that year. With her students at Georgetown she had been studying the motions of 888 relatively nearby stars, following the same methodology she'd used in her master's and doctoral theses — by consulting catalogues.* Now she would be continuing that work, only she would be using a telescope and collecting the evidence herself.

While most astronomers at the time were studying the motions of stars in the interior of the Milky Way galaxy, she went in the other direction — the galactic anticentre, astronomers call it: the stars that lie at greater distance from the galaxy's central bulge than our own star, the Sun. The following year she received an invitation to become the first woman to observe at Mount Palomar, in the mountains northeast of San Diego.† She decided that the time had come to do what she

---

* When the editor of the *Astronomical Journal* said that, as a matter of policy, the resulting paper couldn't list the names of students as authors, Rubin offered to withdraw it. The editor declined her offer and the article appeared with the students' names intact.

† Women had previously not been welcome at either Mount Palomar or its nearby Carnegie Institution sibling, Mount Wilson, ostensibly because the observatories didn't have facilities for both sexes. "This," the astronomer Olin Eggen grandly announced to Rubin on her first tour of Mount Palomar, throwing open a door, "is the famous toilet."

couldn't do as an assistant professor using the increasingly limited resources at Georgetown: become a full-time astronomer.

By now she was living in a quiet, leafy neighborhood near the Department of Terrestrial Magnetism, where she used to meet Gamow in the library. Occasionally she would make the fifteen-minute walk there to visit her friend Bernard Burke and discuss the radio-astronomy analysis he was performing on the rotation of the Milky Way. In December 1964, however, she visited for a different purpose. Even though DTM, founded in 1904, had never had a woman staff member, she walked into Burke's office and asked for a job.

"He couldn't have looked more surprised if I'd asked him to marry me," she told her husband that night.

Once Burke had recovered his composure, he took her to the communal lunchroom and introduced her to his colleagues; she was impressed that one of them, W. Kent Ford, had recently returned from Mount Wilson. Someone encouraged her to go to the blackboard—it was that kind of lunchroom—and talk about her latest work. Before Rubin left DTM that afternoon, Merle Tuve, DTM director and a longtime (since the 1920s) staff scientist, gave her a two-inch-by-two-inch photographic plate and asked her to perform a spectroscopic analysis. After she returned the plate along with her analysis, Tuve phoned to set up an appointment.

She said she could be there in ten minutes.

He said the following week was fine.

She said she could be there in ten minutes.

As was the case at all the Washington labs of the Carnegie Institution, the responsibilities of a staff scientist at DTM included no teaching, no tenure, and infrequent, if any, grant-writing. All it required was the ability to maintain a collegial atmosphere with colleagues and produce meaningful science. In Rubin's case, she had the choice of sharing an office with her friend Bernie Burke* or with W. Kent Ford. Burke was an on-staff radio astronomer, as were Tuve and Kenneth Turner; she noticed that the radio astronomers had commandeered a

* Between her visit to Burke's office in December 1964 and her first day of work on 1 April 1965, Burke took the lunchtime phone call in which he directed Arno Penzias at Bell Labs to Bob Dicke at Princeton.

room on the first floor, thoroughly blanketing a large table with geological layers of charts and other paperwork. Rubin didn't want to immerse herself in their world. She wanted a world of her own, and she figured she was more likely to find it in Ford's office.

Ford was an instrumentalist. He had recently built an image-tube spectrograph—a variation on the standard instrument that records the electromagnetic spectrum from a source of light. His version, however, didn't photograph the light from a distant object. Instead, it converted those faint photons into a fountain of electrons, which in turn sprayed onto a phosphorescent screen, which in turn gave off a vivid glow—and *that* was what his instrument photographed with all the clarity of a "normal" camera. The intensity of the image compensated for the dimness of the distant source of light. As a result, the instrument reduced the exposure time to one-tenth that of an unaided photographic plate. In Ford's new spectrograph—officially the Carnegie Image Tube Spectrograph—Rubin saw the chance to join the hunt for what was then astronomy's hottest prey.

Quasars—short for quasi-stellar radio sources—were extraordinarily powerful pointlike signals, possibly from the farthest depths of space. Their discovery in 1963 provided breathtaking evidence for astronomers that the universe visible in radio waves is not the universe we see with our eyes. And the quasar work that Rubin and Ford did with the new image-tube spectrograph was not unrewarding. Only months after they'd published one of their findings, Jim Peebles was using their data to advance a theoretical exploration of the early universe. Rubin was thrilled. Her research, she marvelled, was contributing to a subject she had never even thought to investigate.

On the whole, though, the two years she spent chasing quasars were burdensome. The field was too crowded, competition for time on the big telescopes favoured more established astronomers from more mainstream institutions, and the pressure to provide data to her non-image-tube-blessed colleagues was crushing. Constantly they insisted on answers even though she wasn't yet sure her answers were right.

This wasn't the way she wanted to do astronomy. She already had enough personal pressures in her life; she didn't need professional pressures, too. So she quit quasars.

She was beginning to realize that in her earlier work she hadn't known what the mainstream was because she herself wasn't working at mainstream institutions. Cornell was no Harvard or Caltech when it came to astronomy; Gamow and de Vaucouleurs weren't the masters of Mount Wilson or Mount Palomar. In those days, though, her outsider status had been inadvertent. Not this time. At least now, she told herself, she knew what the mainstream was; she knew what she was leaving behind. She would shape her observing programme accordingly.

She needed to find a subject she could explore with small telescopes, the kind that generally would be more available to someone of her relatively junior status. She wanted a research program that nobody would care about while she was doing it. But she also wanted it to be work that the community would eventually be glad someone had done.

She found it next door, cosmically speaking: Andromeda, the nearest galaxy that resembles our own.

"Within a galaxy, everything moves," Rubin would write. "In the universe, all galaxies are in motion." Every two minutes "the earth has moved 2500 miles as it orbits the sun; the sun has moved 20,000 miles as it orbits the distant center of our galaxy. In a 70-year lifespan, the sun moves 300,000,000,000 miles. Yet, this vast path is only a tiny arc of a single orbit: it takes 200,000,000 years for the sun to orbit once about the galaxy."

Yet such is the scale of the universe that astronomers don't see galaxies actually rotating. If observers in Andromeda were studying our galaxy — a scenario that Rubin enjoyed imagining — they would see an apparently motionless spiral. So do we when we look at Andromeda. The spectrograph, however, would tell a different story: how much the light from Andromeda had shifted towards the blue or the red end of the electromagnetic spectrum — how fast it was advancing towards or receding from Vera Rubin.

In effect, she had inverted her earlier approach to the image-tube spectrograph. She would still be looking at fainter and fainter objects. But rather than pushing deeper and deeper into space, she would be looking at subtler and subtler details close to home. And she would be doing it in record time.

When the American astronomer Francis G. Pease studied that same galaxy in 1916, he needed eighty-four hours of exposure time over a three-month period to record a spectrum along one axis of the galaxy; the following year he needed seventy-nine hours over a three-month period to record a spectrum along the other axis. Instruments had improved since then, but even by the mid-1960s obtaining a single spectrum of a galaxy still took tens of hours over several nights (assuming that you could even guide the telescope precisely enough and keep the spectrograph stable enough for such a long period, always iffy propositions). Ford's new instrument, however, could reduce the exposure time by 90 percent. Obtaining four to six spectra in one night was routine. In Ford's instrument Rubin saw the potential to measure the rotation motions of Andromeda farther from its central bulge than any astronomer had ever measured on any galaxy before.

Again and again Ford and Rubin made the trip to the two main observatories in Arizona. On occasion their families joined them — Ford had three children, and the two families socialized in Washington — but mostly they went alone. In the dark of the dome, Rubin and Ford would sometimes bump heads — literally — as they each tried to be the one to guide the instrument. In general, though, their competitiveness was restricted to who would spot the first saguaro on the drive south from the Lowell Observatory in Flagstaff. For part of the way on the three-hundred-mile drive — through Phoenix and Tucson to Kitt Peak, with Ford's image tube safely tucked in back — they talked about their children. But mostly, during those days and nights out west, they talked about science.

At the December 1968 AAS meeting Rubin announced that she and Ford had achieved their goal. They had gone farther from the centre in their observations of Andromeda than any other astronomers had gone in observing a galaxy. After Rubin's talk, Rudolph Minkowski, one of the most eminent astronomers of the era, asked her when she and Ford were publishing their paper.

"There are hundreds more regions that we could observe," she said, referring to Andromeda alone. She could gather this sort of data forever. It was beautiful. It was clean. It was unobjectionable. It was what it was.

Sternly, emphatically, Minkowski addressed her. "I think you should publish the paper now."

So she and Ford did. But they knew that before they could submit a formal paper on their research, they would have to address a problem that had bothered them from almost the first night of observing Andromeda.

Going into the darkroom, Rubin had expected to detect the pattern that holds for the planets in our solar system: the farther the planet from the Sun, the slower the orbit—just as Newton's universal law of gravitation predicted. A planet four times as far from the Sun as another planet would be moving at half the velocity. A planet nine times as distant would be moving at one-third the velocity. Pluto is one hundred times as far from the Sun as Mercury, so it should be moving—and does move—at one-tenth the velocity of Mercury. If you plotted this relationship between distance and velocity on a graph—the farther the distance, the slower the velocity—you would get a gradual falling-off, a downward curve.

That's what Rubin and Ford had assumed they would see in plotting the relationship between distance and velocity in the different parts of a galaxy: The farther the stars were from the centre of the galaxy, the slower their velocity would be. That's what astronomers had always done—assumed they would get a downward curve, as if the great mass of stars making up the central bulge in the galaxy affected the wispiest tendrils in the same way that the great mass of the Sun in our solar system affected the wimpiest planet. But those astronomers hadn't actually made those observations because, without the benefit of Ford's spectrograph, they couldn't have. Instead, they drew their assumption as a dotted line. Rubin and Ford, however, had pushed the observations farther than ever, as far as the image-tube spectrograph would allow them, to the farthest edges of the spiral. But they couldn't help noticing that the outermost stars and gas seemed to be whipping around the centre of the galaxy at the same rate as the innermost stars and gas. It was as if Pluto were moving at the same speed as Mercury. Plot the rotation curve of Andromeda, and it wasn't a "curve" at all.

Maybe the gas was interacting with the stars in some way Rubin couldn't imagine. Maybe Andromeda was just an oddball galaxy.

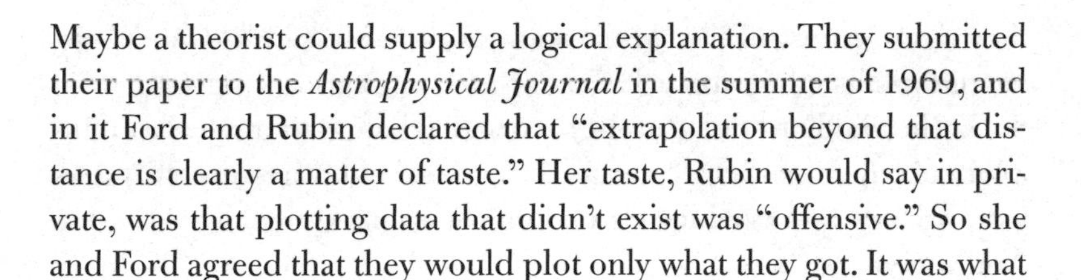

Maybe a theorist could supply a logical explanation. They submitted their paper to the *Astrophysical Journal* in the summer of 1969, and in it Ford and Rubin declared that "extrapolation beyond that distance is clearly a matter of taste." Her taste, Rubin would say in private, was that plotting data that didn't exist was "offensive." So she and Ford agreed that they would plot only what they got. It was what it was.

And what it was, was a flat line.

Shortly after Rubin finished her work on Andromeda, her good friend Morton Roberts, at the National Radio Astronomy Observatory in Charlottesville, Virginia, called to say he was driving over. He had something he wanted to show her.

They met in a basement conference room at DTM, along with a group of three or four other DTM astronomers. Roberts, too, had been studying the rotation curve of Andromeda, except his observations were at radio wavelengths. He placed a copy of the *Hubble Atlas of Galaxies* on the table and opened it to a photograph of Andromeda. Then he laid the plot of his radio observations on the photograph. He had pushed far past the familiar cyclone of stars and gas, far past the point that Ford and Rubin had managed to reach with their optical probes, into a ring of hydrogen gas clouds. But a graduate student from Harvard who was spending some time at DTM, Sandra Faber, seemed unimpressed.

"There's nothing new in this," she said. "It's all part of the same problem. Velocity has never made sense."

She was right. As Rubin herself had shown, velocities of galaxies varied all over the map of the heavens. But for Faber the problem was a given. Unlike Rubin, she'd come of age in a universe that was in motion in more ways than anyone had ever imagined.

"Don't you understand?" Roberts said. "The galaxy has ended, but the velocities are flat." He gestured at the points he'd plotted. "What is the mass out there? What is the matter? There's got to be matter there."

They all stared at the photograph. Here was this beautiful swirl of billions of stars—the kind of majestic image that had captivated astronomers for more than half a century—though that's not where they were looking. They were looking beyond it. Beyond the bulge, beyond the stars, beyond the gas of the spiral arms—beyond all of the light, whether optical or radio. And even though there was nothing to see there, the small group of astronomers understood that they were nonetheless looking at the Andromeda galaxy.

It was what it wasn't.

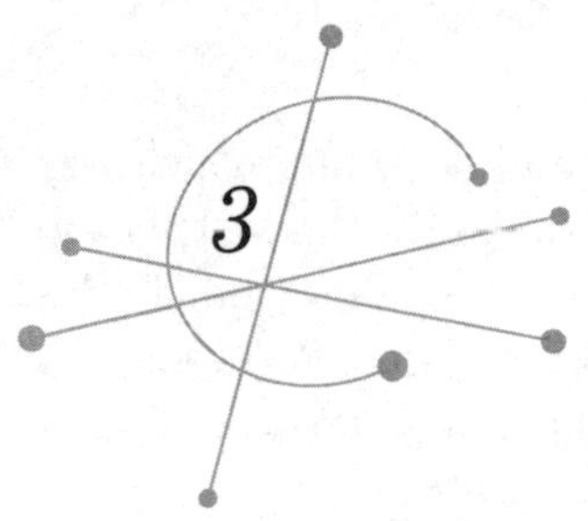


# Choosing Halos

IN THE SUMMER OF 1969, Jim Peebles decided to find out just how simple the universe was.

He had spent the previous academic year at Caltech, and now he and his wife, Alison, were driving back across the country to their home in Princeton. Along the way they stopped at Los Alamos Scientific Laboratory. The lab had invited Peebles to spend a month there as part of a programme to bring outside perspectives into what would otherwise be an insular scientific community in the middle of the New Mexico desert. Los Alamos was where the first atomic bombs were designed: one for the Trinity test, on 16 July 1945, two hundred miles south of Los Alamos, in the arid flatlands outside Alamogordo; then Little Boy, twenty days later, over Hiroshima; then Fat Man, another three days later, over Nagasaki. In 1969, Los Alamos was one of two government facilities (along with Lawrence Livermore National Laboratory, in California) designing nuclear weapons. When Peebles looked around at the supercomputers at the facility, he realized, with characteristic restlessness, that as long as he was there he might as well get some work done.

Forty years earlier Edwin Hubble had arrived at the evidence for an expanding universe by studying the behaviour of galaxies. By tracing that expansion backwards, as if running a film of the outward-flying galaxies in reverse, Georges Lemaître had arrived at the idea of a primeval atom. Peebles hadn't believed that the universe could be that simple, but now he was becoming one of the leading interpreters of a simple universe: homogeneous and isotropic — one that looked the same no matter where you were in it and no matter which way you looked. After Penzias and Wilson, as well as Peebles and his colleagues at Princeton, had found evidence for what those initial "Primeval Fireball"* conditions might actually have been, Peebles began using that knowledge to refine his understanding of the expansion itself. It was as if he were running the film of the history of the universe again, only instead of rewinding it to the beginning he would be running it forwards to today.

Now *that* was a movie he wouldn't mind seeing.

The computer he would be using at Los Alamos — a CDC 3600 — was many magnitudes more powerful than any he could have found on a university campus, and he wouldn't even have to tap the Princeton Physics Department's research funds. He could run the computer as long as he needed — all night, even all weekend. And Peebles could do so even though Los Alamos was in high Cold War mode and he wasn't a U.S. citizen. Peebles had emigrated from Manitoba only eleven years earlier. He was a Canadian citizen — officially an alien. Yet apparently the work he was doing seemed either so primitive or so esoteric — or his demeanour so unthreatening; his reputation so established; his computational skills so (relatively) undeveloped — that his entire security detail consisted of a secretary who sat to one side, knitting.

Peebles would be performing what scientists call an *N*-body simulation. Take a number — *N* — of points, programme them to interact according to whatever properties you want, and see how the action unfolds. In this case, Peebles would be taking 300 points and treating each as if it were a galaxy in one particular part of the universe — the Coma Cluster, the closest and most-studied galaxy cluster. He would assign each galaxy a position and velocity based on rough observations of real galaxies in the cluster, and he would teach the computer the law of universal gravitation. And then he'd let the model do whatever galaxies interacting gravitationally in an expanding universe do over billions of years.

He'd already been thinking about how clusters develop, and he'd

---

* The name the Physics Department in Princeton favoured, from a suggestion by John Archibald Wheeler.

done some cluster calculations during his time at Caltech. Now he took that initial research and converted it into a computer program. Then he punched the holes in the 7⅜-by-3¼-inch computer cards himself, stacked them in the metal feeder, and ran a simulation. At the end of the simulation the points had moved a bit. He transferred the image to a frame of 35-millimetre film, and then he ran the next simulation, using the galaxies' positions at the end of the previous simulation as a starting point. Again, over a period equivalent to millions of years, the galaxies shifted slightly. When Peebles had enough frames, he ran them together, loaded the film in a projector, and sat back.

The universe swirled to life. Galaxies moved outwards, following the flow of the Hubble expansion. But then they didn't. They slowed, also moving under the influence of their mutual gravitational attractions, and they continued to slow, until they stopped following the flow of the expansion and began to fall back on themselves. Smaller galaxies clumped toward the nearest larger galaxies, and those growing clumps clumped with other clumps. The more the galaxies clustered, the more the galaxies clustered.

Simple, sort of.

In one form or another, the question of how simple the universe was had come to occupy Peebles ever since the discovery, four years earlier, of the cosmic microwave background at a temperature that roughly matched his prediction. Despite his earlier mistrust of cosmology, he now felt that the field had probably possessed the requisite makings of a science since the early 1930s. Edwin Hubble had acquired a set of measurements: the direct correlation between galaxies' distances and redshifts. Georges Lemaître and Aleksandr Friedman had attached a theoretical interpretation to those observations: a universe expanding from a Big Bang. And there was no shortage of open issues to drive further research. The assumptions that the universe would look the same on the largest scale (that it was homogeneous) no matter which way you looked (that it was isotropic) were not assumptions that Peebles would have made. And maybe his prejudice against simplistic assumptions had blinded him to the scientific possibilities of cosmology. But his attitude had begun to soften even before he heard of Penzias and Wilson's 3 K detection. As soon as he had finished his calculations for the cosmic microwave background's temperature—a prediction you could actually put to a test—he understood that he was going to have to take cosmology seriously.

At once he and Dicke began collaborating on a major paper, "Gravitation and Space Science," which they sent to *Space Science Reviews* in early March 1965, right around the time of the fateful phone call from Penzias. (They added a note in proof about the Penzias and Wilson detection as well as the two upcoming papers in the *Astrophysical Journal.*) Dicke handled the physics section of the paper, Peebles the cosmology. As a sort of belated rejoinder to the Rochester professor who a quarter of a century earlier had told Dicke that physics and general relativity had nothing to do with each other, they wrote in the introduction: "While in a limited sense gravitation is of no great importance to a physicist, this is much too naïve an interpretation." In the first paragraph of the "Cosmology" section Peebles expanded on that philosophy. For physicists, he wrote, cosmology doesn't satisfy just "the obvious interest" in the origins of the universe; "we need cosmology as a basis of any complete theory of the galaxies, or for that matter, of the solar system."

If you wanted to understand specific problems concerning the evolution and structure of the universe—the clustering of galaxies, for instance—then you had to abandon any residual "island universe" assumptions. You had to learn to think about the universe not only as a collection of individual galaxies but as the sum of its galaxies—a single unit, a whole. You had to keep in mind that while the whole—the universe—was expanding, its parts—the galaxies—were evolving. "The moral of this section," Peebles offered in conclusion, twelve pages later, "is the unity of the universe."

Even as he was collaborating with Dicke on "Gravitation and Space Science," Peebles had begun writing an investigation into how the primeval conditions of the universe might develop into galaxies. Like the paper he wrote with Dicke, it went out in the post in early March; unlike the other paper, it would later receive major revisions to accommodate the 3 K discovery. By the time this paper ran in the *Astrophysical Journal,* in November 1965, the Princeton and Bell Labs articles had appeared in print, and many of Peebles's peers were rapidly moving past the Why-take-cosmology-seriously? and into the How-exactly-does-this-cosmology-thing-work? phase.

Peebles showed them how: Look at the maths. Look at the prediction; then look at the observation. See how they match? During one presentation at an American Astronomical Society meeting, Peebles was pacing back and forth before a blackboard, taking long strides and waving his chalk, when an astrophysicist in the audience called out, "One can make any point at all with a little slapdash arithmetic!"

Peebles spun toward the voice, smiling broadly. "My arithmetic may seem slapdash," he said with a flourish, never breaking stride, "but I can assure you it is impeccable."*

"The radiation," Peebles explained in one of his many papers during this period, "performs the great service of defining the epoch at which the galaxies can start to form." That epoch occurred when the temperature of the primeval fireball fell below 4000 K. At that point the electrons and protons that had been ricocheting independently since the first instants of the universe recombined to form atoms of matter. This matter now took on a "life" of its own and decoupled from the radiation—the fossil radiation that survived today as the cosmic microwave background. And although that background seemed to be as uniform—as homogeneous—as theory had predicted, it couldn't be *entirely* uniform. It had to contain the irregularities, or inhomogeneities, that identified the concentrations of matter that existed at the moment that matter and radiation decoupled and went their separate ways—the matter that through gravitational interactions would have grown into the large-scale "distributions of mass and size" that we see today: "galaxies, and clusters of galaxies, and the material within galaxies," including us.

The universe *was* simple. It just couldn't be *perfectly* simple. Yet to the radio telescopes at the time, the microwave background was absolutely uniform; it lacked the inhomogeneities that had to be there in order for us to be here. Eventually, astronomers or physicists wanting to test the Big Bang theory would have to develop instruments sensitive enough to detect those subtle irregularities in the background. In the meantime Peebles would proceed, as always, with caution.

Having already made the movie of the universe, Peebles now wrote the book. In the fall semester of 1969, after returning from Caltech to Princeton, Peebles taught a graduate course on cosmology. His col-

---

* The astrophysicist wrote a note of apology to Peebles that evening.

league John Archibald Wheeler—legendary theorist, Princeton fixture since 1938, longtime collaborator of Einstein's—suggested that Peebles use the course as the basis for a text, but Peebles demurred. So Wheeler began showing up at Peebles's lectures. Peebles would be pacing at the blackboard, enthusing in his usual wingspan-enhanced fashion. Wheeler would be sitting in the back, taking notes. Thoroughly unnerving Peebles. After the lecture, it only got worse. Wheeler would present Peebles with the notes, written in perfect penmanship. The "blackmail"—Peebles's word—worked.

Wheeler, he soon saw, was right. If cosmology was in fact in the process of graduating from speculation to science, from metaphysics to physics, then it deserved a textbook. It *needed* a textbook, if a new generation was going to investigate cosmology properly. The cosmology texts still in use were the ones that Peebles had consulted as a graduate student while preparing for his general examinations; they were cobwebbed with decades-old theory of the kind that valued the simplicity of the maths over the relationship to observations, if only because, before the advent of radio astronomy, those observations couldn't be made.

Peebles conceived *Physical Cosmology* as the first full-length examination of the physics of the early universe. Such was the state of knowledge that he could explain the entire field in 282 pages. "The great goal now," Peebles wrote in the introduction, "is to become more familiar with the Universe, to learn whether any of these pictures may be a reasonable approximation, and if so how the approximation may be improved." He kept in mind a conversation he'd had with the physicist Philip Morrison shortly after the discovery of the radiation. The two of them were standing in a crowded room. "You measure the level of noise in this room," Morrison said, "and convert that into a thermal temperature and you'll get an absurd answer. How do you know you aren't doing the same thing here?" Peebles's answer: He didn't know. Nobody knew. Although the discovery of the 3 K radiation had provided his career with a new trajectory and inspired the book, he discussed the microwave background only as a "candidate" for the primeval fireball.

Peebles wouldn't even venture into the area of research that had been occupying so much of his time recently—"the very broad topic of galaxy formation, and the presumably related task of understanding irregularities of all sorts in an expanding universe." He could describe the behaviour of clusters of galaxies but not the behaviour of galaxies themselves. Some papers he was seeing—for instance, Mort Roberts's radio observations of Andromeda—were even indicating that the rotation curves of individual galaxies might be flat. Even though he'd included galaxy formation in the lectures themselves, he mostly omitted the topic from his book. As provisional as much of the cosmological physics in the rest of the book might be, the knowledge of individual galaxy formation was so raw that he decided it wasn't yet worth committing to a bound volume.

So Peebles was hardly surprised when a Princeton colleague, the astronomer Jeremiah Ostriker, stopped by his office to say that he couldn't make sense of the behaviour of the Milky Way. Ostriker had seen the *N*-body simulations that Peebles loved to demonstrate, and he'd consulted with Peebles on the resulting paper. (With the help of a graduate student, Peebles had gotten the *N* up to 2,000.) Ostriker had been working on rotating celestial objects since he was a graduate student at Cambridge; he had written his thesis on rotating stars. Scientists had known since the nineteenth century that if you rotated an initially spherical liquid drop it would become oblate, increasingly so, and eventually compress into a bar shape. Ostriker had treated stars as liquid drops—as compressible objects—and found that they, too, would become oblate over time. Recently, he told Peebles, he had looked at a rendering of the Milky Way—a flat disk like the other spiral galaxies that astronomers had been collecting by the thousands. He could see at a glance that it should have become bar-shaped or broken up into two galaxies after one rotation. Yet by now the Milky Way was old enough to have completed a dozen rotations.

"There's something wrong here," Ostriker said.

Peebles agreed. He began work on creating an *N*-body simulation for the Milky Way. He laid the points into a spiral shape and set it rotating. Sure enough, it wobbled catastrophically during its first 200-million-year rotation. He and Ostriker would need something else to stabilize it: a surrounding mass of something to hold it together gravitationally. Something you couldn't see with a telescope—at least not yet—but something that had to be there. Something that Peebles and Ostriker would now add to the computer program.

For the first simulation of the rotation of the galaxy with this missing component, the amount of mass wouldn't matter; it would simply serve as a basis for further comparison. Ostriker and Peebles would surround the visible galaxy with this mass and see what happened. If they ran the simulation and the rotating disk stabilized, they'd shrink the halo and keep shrinking it until the disk destabilized. If they ran the simulation and the rotating disk didn't stabilize, they'd expand the halo and keep expanding it until the disk did.

Peebles shrugged. "Just choose a halo."

They did — a large one, one that swallowed a huge portion of the visible galaxy. Once again the Milky Way wobbled and gyrated. So they tried a larger halo. Another unstable galaxy. Another halo, another unstable galaxy. And another, and another.

This result wasn't entirely surprising to Peebles. The problem of "missing mass" had been shadowing astronomy for decades, for almost as long as astronomers had known of the existence of galaxies. But the problem had always related to *clusters* of galaxies. In 1933 the Swiss-born astrophysicist Fritz Zwicky, working at Caltech, studied eight galaxies in the Coma Cluster, comparing the mass he derived from their velocities relative to one another with the mass he expected just judging from appearances. His conclusion was that the density of mass had to be four hundred times as large as what the luminosity alone suggested.* If astronomers couldn't resolve this discrepancy, he wrote in a Swiss journal, "we would arrive at the astonishing conclusion" that the density of luminous matter in Coma must be minuscule compared with the density of some sort of *dunkle* — or dark — *Materie* — matter. Three years later, the astronomer Sinclair Smith published an article in the *Astrophysical Journal* about a similar pattern he'd noticed in the Virgo Cluster, suggesting the presence of "a great mass of internebular material within the cluster." That same year, Edwin Hubble addressed the problem in his landmark book *The Realm of the Nebulae:* "The discrepancy seems to be real and is important."

"This discrepancy is so great," Zwicky wrote in 1937, "that a further analysis of the problem is in order."

---

* Later reduced by other astronomers to fifty times — but still . . .

Further analyses came, but only sporadically and inconclusively. In science, progress often follows a self-fulfilling logic: You work on the problems that either have the best chance of yielding conclusions or are most in need of them. Astronomy in the post-Hubble, galaxies-aplenty era was lousy with such problems. The motions of poorly understood objects (galaxies) in possibly coincidental formations (clusters) was not one of them. Peebles himself had regarded the missing-mass problem as one of those topics you discussed only if you were shooting the breeze during a coffee break, like the question of what came before the universe.

The rise of cosmology as a real science in the late 1960s, however, suddenly made the missing-mass problem more pressing. If you considered the evolution of the universe on the largest scales — as Peebles had, in *Physical Cosmology* — you couldn't exactly ignore the behaviour of the largest structures in the universe, galaxy clusters. His conclusion there, however, could only echo Zwicky and Hubble's plea more than three decades earlier: "We urgently need comparable data."*

Now, though, he was seeing a similar problem in his simulations of individual galaxies, and he had begun seeing the same pattern in papers claiming flat rotation curves in galaxy after galaxy. What if missing mass wasn't a problem just in galaxy clusters? What if it was a problem in individual galaxies, too? What if it was *the same problem?*

Peebles and Ostriker kept giving their galaxy a larger and larger halo. Only when they had simulated an invisible halo with roughly the same mass as the visible parts of the galaxy — the disk of stars and gas, of central bulge and spiral arms — did the system stabilize. In 1973 Ostriker and Peebles published a paper arguing that "the halo masses of our Galaxy and of other spiral galaxies *exterior* to the observed disks may be extremely large."

Over the following year, working with a postdoc, they wrote a second paper on the subject. Instead of running hypothetical models on

* In a 1969 paper that he adapted from a talk he'd given two years earlier, Peebles mentioned that the density of matter in galaxies "could be augmented by dark matter" — perhaps the first use of the term since Zwicky. It was, however, an anomalous usage; Peebles otherwise adopted the industry standard "missing mass."

their computers, they analysed observations that astronomers had already made. They examined the data on individual galaxies. They examined the data from binary galaxies—pairs of galaxies where each galaxy closely interacts gravitationally with the other. They examined the data from satellites of galaxies—dwarf elliptical galaxies that orbit large spiral galaxies. And when they were done, they collected all the data into one comprehensive table, compiled their analysis, and made, in the opening sentence of a second paper, a deceptively simple declaration: "There are reasons, increasing in number and quality, to believe that the masses of ordinary galaxies may have been underestimated by a factor of 10 or more."

Just brilliant.

Vera Rubin read the opening sentence of that second paper and recognized the kind of breadth of vision and distillation of ideas that could redefine a field. And she wasn't alone. The two papers created a sensation, though not the kind that Peebles and Ostriker might have hoped for. People were angry. And while Peebles hardly noticed (observers, upset with theorists?—so be it!), Ostriker felt such intense hostility that he had to wonder: Were most astronomers even reading what was on the page? Or were they just having a visceral reaction to the possibility that what they had been studying this whole time was only 10 percent of what was actually out there? Either way, most astronomers still weren't in the habit of thinking about the relationship between gravity and galaxies.

Vera Rubin was. After she and Kent Ford completed their paper on Andromeda in 1969, she had turned her attention to the question that had motivated her master's thesis: Did the universe rotate? Or, in more mature terms, did the distribution and velocity of galaxies suggest a lack of uniformity beyond the local universe—on the kind of scale that could make the universe a little less simple?

Twenty years had passed since that rancorous AAS meeting in Haverford. Rubin was no longer an unknown neophyte relying on the research of real astronomers. She had made a name for herself (and it wasn't Vera Hubin); her early work had been vindicated. Gérard de Vaucouleurs, Rubin's constant correspondent during the 1950s, had published several papers over the years showing results similar to her own. By the 1970s the pattern of non-uniform distribution of galaxies on a relatively local scale was, as she wrote in a paper during this period, "well discussed"; her fellow astronomers had adjusted themselves to the evidence that some galaxies were clustering even as the universe as a whole was expanding. The general assumption, however, was that at greater distances the universe would be the same in every direction — that any departures from homogeneity and isotropy were local, and that on larger scales the galaxies would adopt a more uniform distribution. The question Ford and Rubin (and her daughter, Judith, a student at Radcliffe College) would address was: Did galaxies really behave this way?

"The results," they wrote in 1973, while still collecting their data, "are so striking that we wish to present a preliminary account."

Once again Rubin found that galaxies exhibited not just the recessional motion of the expansion but peculiar motions. In this case, a group of local galaxies seemed to be racing together towards one part of the sky. And once again much of the community rejected the conclusion. The Rubin-Ford effect, as it became known, was the subject of virulent arguments at conferences. Prominent astronomers begged Rubin to drop the line of research before she ruined her career. But she and Ford pushed their observing mission to the end and, in 1976, published the complete set of data in two papers that they felt established the Rubin-Ford effect as real.

As usual, Rubin didn't like the controversy. She didn't like everyone challenging her on every number. She didn't want to have to defend her data. She didn't want to have to defend the universe. She would say that she wasn't "smart enough" to know why the universe was the way it was: "I could design a woman's plumbing. But the universe, I couldn't do it." The universe was what it was. And she was who she was. Shortly after publishing the papers on the Rubin-Ford effect, she attended a Yale conference on galaxies; above the entrance hung a giant banner: ASTRONOMERS. She walked under it. "Okay," she thought wryly. "Now I'm an astronomer."

Besides, she and Ford had something else to pursue. They had seen a continuation of the phenomenon that they had noted in their 1970 paper on Andromeda and that Mort Roberts had shown them in his radio observations of the same galaxy. In their observations of galaxies that led to the Rubin-Ford effect, they looked at galaxies far more distant than Andromeda, and therefore far smaller from the point of view of an observer on Earth. They could see the galaxy in one gulp. In the end they studied sixty galaxies, and even though Rubin was using the spectroscope to measure the motions of entire galaxies, the rotation curves showed up anyway, a shadowy residual effect. These rotation curves looked flat too, just like Andromeda's, at least at a glance. Would they still look flat under more rigorous, more focused examination? Rubin decided to do what an observer does: more observing.

For their 1970 paper she and Ford had pushed as far out to the periphery of Andromeda as 1960s technology allowed. In 1974 a new 4-metre telescope opened at Kitt Peak, twice the diameter and therefore four times the surface area of the one they had used in observing Andromeda. The combination of Ford's spectrograph and a significantly larger telescope would allow them to take their study of galaxies both deeper into the universe and farther along the arms of the spirals. In 1978 Ford and Rubin published the rotation curves for eight more galaxies: all flat.

Once again radio astronomers were getting the same results. Mort Roberts kept pushing along a ring of hydrogen gas clouds that lay beyond the visible swirl of stars and gas. In 1975 Roberts and a collaborator found that even there, half the length of Andromeda beyond what previous generations had unthinkingly assumed was the galaxy in its entirety, the rotation wasn't tapering off. It was essentially flat, as if even at this great distance the galaxy was still spinning at a seemingly suicidal rate. A 1978 survey using the same method found the same shape for the curves in twenty-two of twenty-five other galaxies: flat.

Rubin had gotten her wish. The data spoke with one voice, and it spoke clearly: Galaxies were living fast but not dying young. Observers and theorists could question the evidence and double-check the methodologies, as they should and did. Some suggested that radio observations were of necessity indistinct; they covered too much of the sky to provide reliable data. Some suggested that optical data like Rubin's suffered from a bias; she was looking only at high-luminosity galaxies because they were the easiest to find, and maybe their masses were anomalous. Some suggested that elliptical galaxies wouldn't show the same flat rotation curves as spiral ones. But even the most ardent critics were finding it difficult to quarrel with the uniformity of the data. Plot after plot from astronomer after astronomer in journal after journal — all a sceptic had to do was look at the rotation curves. You could see where the sources of light were. You could see where the motions of the galaxy said the mass should be. And you could see that the two didn't match.

In 1979, in an article in the *Annual Review of Astronomy and Astrophysics,* two astronomers — including Sandra Faber, who had been unimpressed by Mort Roberts's flat rotation curve when she was a graduate student visiting DTM — looked at all the evidence they could gather. "Is there more to a galaxy than meets the eye (or can be seen on a photograph)?" they wrote in the opening sentence. Their conclusion, forty-seven pages of exhaustive analysis later: "After reviewing all the evidence, it is our opinion that the case for invisible mass in the Universe is very strong and getting stronger."

A couple of years earlier, Rubin had come away from the Yale conference on galaxies with the impression that, as she wrote, "many astronomers hoped that dark matter might be avoided." Now, the publication of Faber and Jay Gallagher's comprehensive argument left most astronomers agreeing that their field had a problem with "missing mass" — though this term increasingly seemed like a misnomer. After all, the problem wasn't that astronomers didn't know where the mass was. They did. It was in the halo — or at least in a "massive envelope," the term that Faber and Gallagher adopted in an effort to be "neutral" as to the shape. The problem for astronomers was that they couldn't see it. Not with their eyes, not with a traditional optical telescope, not with a telescope that could see in any wavelength of light. In which case, the mass wasn't "missing" at all. It was just — to borrow the term that Zwicky had used in 1933 — *dunkle:* dark.

"Nobody ever told us that all matter radiated," Vera Rubin liked to say. "We just assumed that it did." Her tone, like the reaction in Dicke's office on the day he got the phone call about the detection at Bell Labs, was not one of disappointment. Instead, she felt that by "recognizing that they study only the 5 or 10 percent of the universe which is luminous," astronomers "can approach their tasks with some amusement."

The joke was on us. In 1609 Galileo had discovered that looking farther into space than what he could see with the naked eye led to seeing more of the universe. Since the middle of the twentieth century, astronomers had discovered that looking farther along the electromagnetic spectrum than what they could see with an optical telescope led to seeing even *more* of the universe—including the echo of its origins. And now, if you were Vera Rubin, you could look up from your desk and gaze at the giant photograph of Andromeda that you'd hung on the ceiling, and you could ask, with greater sophistication than a ten-year-old leaning on a bedroom windowsill but with the same insatiable wonder: How could you possibly see farther than the electromagnetic spectrum—farther than seeing itself?

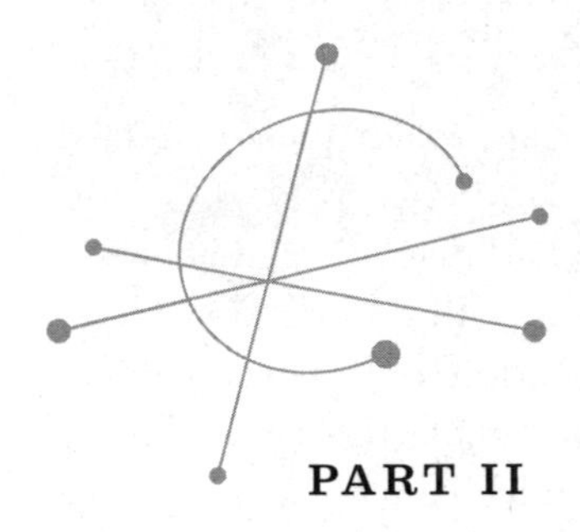

PART II

# Lo and Behold

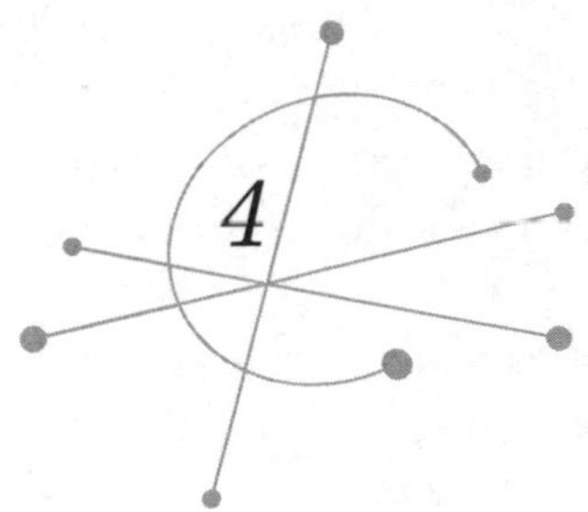


# Getting in the Game

THE WEIGHT OF the universe. The shape of the universe. The fate of the universe.

They talked about it in those terms. They used this giddy language in proposals to solicit funding. They used it in a brochure to recruit graduate students. They used it with the other members of their collaboration as they all told themselves that they were the ones who were finally going to solve some of the most profound mysteries of cosmology — of civilization itself. They also used this language when they needed to reassure themselves that they weren't rebuilding Babel or emulating Icarus, that their experiment was an exercise not in hubris but in science.

Okay, maybe a little hubris. Saul Perlmutter wasn't a born astronomer. He hadn't collected telescope parts as a child, hadn't sketched the motions of the night sky, hadn't dreamed of solitary vigils on mountaintops, just him and the heavens. Carl Pennypacker wasn't a born astronomer either, though at least his PhD in physics was on a related topic, infrared astronomy. And the other members of their team weren't astronomers. None of them had come to Lawrence Berkeley National Laboratory to do astronomy; astronomy wasn't what LBL usually did. Still, they had reason to think they were in the right place at the right time.

The right place, because LBL and the University of California, Berkeley, had just won a government competition to establish a major new research centre. The name on the proposal was the Center for Particle Astrophysics, though because the titular "particle" was dark matter, they could have called it the Dark Matter Center—and, as the first director of the centre once said, they probably would have if they'd thought of it.

And the right time, because by the 1980s scientists could proceed under the assumption that they were in possession of the middle and the beginning of the cosmic narrative. They knew that their protagonist—the universe—was expanding. They had a reasonable explanation for how it had gotten to this point in the story—the Big Bang. Now they could ask themselves: What will become of Our Hero?

Did the universe contain enough matter to slow the expansion so much that one day it would stretch as far as it could, stop, and reverse itself, like the trajectory of a tossed ball returning to Earth? In such a universe, space would be finite, curving back on itself, like a globe.

Or did the universe contain so little matter that the expansion would never stop but go on and on, like a rocket leaving Earth's atmosphere? In this kind of universe, space would be infinite, curving away from itself, like a saddle.

Or did the universe contain just enough matter to slow the expansion so that it would eventually come to a virtual halt? In this universe, space would be infinite and flat.

Borrowing from the Big Bang example, astronomers gave the first options the cheerfully inadequate names Big Crunch (too much matter) and Big Chill (too little matter); the third option was the Goldilocks universe (just right). From only one measurement, astronomers could determine the weight, the shape, and the fate of the universe.

Before the 1980s, astronomers had certainly known that the amount of matter in the universe would have an effect on the universe's rate of expansion. What they hadn't known was that they had been missing 90 percent or more of the matter. The possible cosmological implications of this realization had been evident from the start. "Not until we learn the characteristics and the spatial distribution of the dark matter," Vera Rubin had written in *Science* shortly after the idea gained widespread acceptance, "can we predict whether the universe is of high density, so that the expansion will ultimately be halted and the universe will start to contract, or of low density, and so that the expansion will go on forever."

Now Perlmutter and Pennypacker set out to make that measurement. They recognized that writing the closing chapter in the story of the universe would be challenging in the extreme, and they figured they would be done in, oh, three years.

The question of how the universe will end was as old as civilization, but the difference now was that scientists might be able to go out and make the crucial measurement. Because the discovery of the 3 K temperature had matched a prediction from the Big Bang theory, it had taught astronomers what to think about cosmology: It just might be a science after all. But the 3 K discovery also taught them *how* to think about cosmology: If you want to understand the history and structure of the universe—if you want to do cosmology—you have to do what Bob Dicke and Jim Peebles had been urging even before the discovery of the cosmic microwave background: think about gravity on the scale of the universe.

Not that astronomers had been altogether ignoring the relationship between gravity and the universe. Much of modern physics and all of modern astronomy had arisen from Newton's epic struggles to derive a law of gravity that was universal. In his *Principia,* published in 1687, Newton met Plato's challenge to find the calculations on paper that matched the motions in the heavens. The telescope had given astronomers the physical tool to chronicle more and more of those motions. But it was Newton's math that had given them the intellectual tool to make sense of them. The law of universal gravitation was what made cosmology-as-science possible.

Yet it also made cosmology-as-science problematic. A syllogism (of sorts): One, the universe is full of matter; two, matter attracts other matter through gravity; therefore, the universe must be collapsing. So why wasn't it?

This was the question that the cleric Richard Bentley posed to Newton in 1692 while preparing a series of lectures on faith, reason, and the just-published *Principia.* Newton acknowledged that his argument required "that all the particles in an infinite space should be so accurately poised one among another as to stand still in a perfect equilibrium. For I reckon this as hard as to make not one needle only but an infinite number of them (so many as there are particles in an infinite space) stand accurately poised upon their points." How was such an equilibrium possible? In a later edition of the *Principia,*

Newton appended a General Scholium in which he postulated an answer—the foresight of God: "And so that the systems of the fixed stars will not fall upon one another as a result of their gravity, he has placed them at immense distances from one another."

What made cosmology scientifically suspect for investigators of nature wasn't just this invocation of a supernatural cause—a cause that was, literally, beyond nature. The problem was the effect. Or, more accurately, it was the *absence* of an effect. Newton's physics was all cause-and-effect, matter-and-motion. Yet what he was proposing in this one instance was a *lack* of gravitational interaction among the bodies of the cosmos. Having conceived of gravity as action at a distance, Newton was now suggesting the need for *inaction* at a distance.

Over the following decades and centuries, the more that astronomers discovered about the system of "fixed stars"—that the stars aren't fixed at all but are in motion relative to one another, and that the entire system of unfixed stars, our galaxy, rotates around a common centre—the less satisfying was the explanation of inaction at great distances.

Einstein made subtle adjustments to Newton's theory of gravity. And in his 1916 theory of general relativity, he presented calculations on paper that matched the motions in the heavens slightly more accurately than Newton's. Yet he, too, had to account for a universe that, as was evident in "the small velocities of the stars," wasn't collapsing of its own weight. In his 1917 paper "Cosmological Considerations on the General Theory of Relativity," he inserted a fudge factor in his equation—the Greek symbol lambda, "at present unknown"—to represent whatever it was that was keeping the universe from collapsing. Like Newton, he feigned no hypotheses as to what that something might be. It was just . . . lambda. But then, little more than a decade later, came Hubble's universe, and with it an elegant and unforeseen solution to the lack-of-collapse conundrum: The reason the universe wasn't collapsing of its own weight was that it was expanding.

Newton hadn't needed God, and Einstein hadn't needed lambda. In 1931 Einstein travelled from Germany to the Mount Wilson Observatory in the mountains northeast of Pasadena and visited Hubble.

After reviewing the expansion data for himself, Einstein discarded his fudge factor. In retrospect, physicists of a philosophical bent came to realize, the problem with cosmology hadn't been a supernatural cause (God). And it hadn't been an illogical effect (inaction at a distance). It had been the unthinking assumption behind the syllogism, the premise of the whole cosmology-as-science debate: that the universe was static.

If you took the universe at face value, as even Einstein did, you would have unthinkingly assumed it was, on the whole, unchanging over time. But the universe (yet again) wasn't what it appeared to be. It wasn't static. It was expanding, and that expansion was outracing the effects of gravity—for now, anyway.

But what about over time? A new syllogism presented itself: One, the universe is expanding; two, the universe is full of matter attracting other matter through gravity; therefore, the expansion must be slowing down. The lingering challenge to aspiring cosmologists was no longer *Why wasn't the universe collapsing?* It was *Will it collapse?*

Ever since Hubble's discovery of evidence for an expanding universe, astronomers had known how to measure how much the expansion was slowing down, at least in principle. Hubble had used Henrietta Swan Leavitt's period-luminosity relation for Cepheid variable stars to determine distances to nearby galaxies. And he had used the redshifts for those galaxies as equivalent to their velocities as they moved away from us. When he graphed those distances against those velocities, he concluded they were directly proportional to each other: the greater the distance, the greater the velocity. The farther the galaxy, the faster it was receding. This one-to-one relationship showed itself as a straight line on a 45-degree angle. If the universe were expanding at a constant rate, that straight line would continue as far as the telescope could see.

But the universe is full of matter, and matter is tugging gravitationally on other matter, so the expansion can't be uniform. At some far reach of the Hubble relation, the galaxies would have to deviate from the straight line. The graph of their points would begin to gently curve toward brighter luminosity. How much they deviated from the straight line would tell you how much brighter they were at that particular redshift than they would be if the universe were expanding at a constant rate. And how much brighter they were — how much nearer they were — would tell you how much the expansion was slowing.

To make that distinction, astronomers would need to continue to graph distance against velocity. For the velocity axis, they could still use redshift. For distance, however, they had a problem. Cepheid variables are visible only in relatively nearby galaxies. For distant observations astronomers would need another source of light that had a standard luminosity, celestial objects they could plug into Newton's inverse-square law as if they were so many candles at greater and greater distances.

And ever since Hubble's discovery of evidence for an expanding universe, astronomers had also known about the standard-candle candidate that the Berkeley team would decide to pursue. In 1932, the British physicist James Chadwick discovered the neutron, an uncharged particle that complements the positively charged proton and the negatively charged electron. In 1934, the same Caltech astrophysicist who had recently suggested that galaxy clusters might be full of dark matter, Fritz Zwicky, collaborated with the Mount Wilson astronomer Walter Baade on a calculation showing that, under certain conditions, the core of a star could undergo a chain of nuclear reactions and collapse. The implosion would race inwards at 40,000 miles per second, creating an enormous shock wave and blowing off the outer layers of the star. Baade and Zwicky found that the surviving ultracompact core of the star would consist of Chadwick's neutrons, weighing 6 million short tons to the cubic inch and measuring no more than 60 miles in diameter.

Astronomers had already identified a class of stars that suddenly flared brighter, then dimmed, and they had named this phenomenon "nova," for "new star" (because its sudden brightening might suggest it was "new" to us). Baade and Zwicky decided that their exploding star deserved a classification all its own: "super-nova."

Almost at once, Zwicky initiated a search for supernovae. He helped design a 460-millimetre telescope that became the first astronomical instrument in use on Mount Palomar, and soon newspapers and magazines across the US were keeping a running tab of how many "star suicides" his survey had discovered. Baade, meanwhile, suggested that because supernovae seemed to arise from the same class of objects, they might serve as standard candles, though he cautioned in a 1938 paper that "probably a number of years must elapse before better data will be at our disposal."

The wait turned out to be half a century. In 1988, the National Science Foundation awarded the University of California, Berkeley, 3.7 million pounds over five years to establish the Center for Particle Astrophysics. The centre would take multiple approaches to the mystery of dark matter. One was to try to detect particles of dark matter in the laboratory. Another looked for signs of dark matter in the cosmic microwave background. A third approach explored dark matter through theory. And another group would try to determine how much matter was out there, dark or otherwise, by using supernovae as standard candles.

The scepticism was sure to be high: physicists doing astronomy? Pennypacker and Perlmutter knew they would eventually have to convince the astronomy community, as insular and guarded as any in science, that physicists working at a particle physics institution could be capable in their line of work. But first they were going to have to convince themselves.

Lawrence Berkeley National Laboratory had invented accelerator particle physics as the world had come to know it. In the late 1920s the physicist who would become the lab's namesake, Ernest Lawrence, conceived of an accelerator that shot particles not in straight lines, as linear accelerators did, but in circles. Strategically placed magnets would deflect the particles just enough to prod them to follow the closed curve, around and around, faster and faster, to higher and higher levels of energy. Lawrence's first "proton merry-go-round" — or cyclotron — was 13 centimetres in diameter, small enough to fit fit inside any room bigger than a broom closet in the physics building on campus. In 1931 he'd moved his operations into an abandoned building, the former Civil Engineering Testing Laboratory — the first official site of the Berkeley Lab "Radiation Laboratory." By 1940, a version of the cyclotron had reached a diameter of 4.7 metres, and the experiment had outgrown the Rad Lab. Lawrence secured from the university a promontory above the campus. But his legacy wasn't just the complex of buildings that over the decades would come to line Cyclotron Road. It was all the particle accelerators in various places around the world that circle underground, miles-long snakes devouring their tails.

What fun was that? You could be a cog in the biggest wheel in the history of the planet, you could even be the most important cog, but you'd still be a cog—assuming you lived long enough for the next generation of accelerators to come online. Luis Alvarez had been a key contributor to the construction of the Bevatron, a proton accelerator with a 122-metre circumference that opened at LBL in 1956. But his was not the kind of mind that aspired to cogdom.

On his own he used principles of physics in a frame-by-frame analysis of the Zapruder film of the assassination of President Kennedy to see whether the "one bullet, one gunman" hypothesis was tenable. (It was.) He probed an Egyptian pyramid with cosmic rays to learn whether it held secret passages. (It didn't.) He wanted to mount a "High Altitude Particle Physics Experiment," but the LBL director at the time refused to give him access to lab funds. Not because of what the experiment would be doing—particle physics was what LBL did. Rather, the problem was how HAPPE would be doing it: aboard a balloon. "We are an *accelerator* lab," the lab director Edwin McMillan, a Nobel laureate in Physics, told Alvarez. "If we stop doing accelerator physics, our funding will disappear."

For Alvarez, turning your back on a particle physics experiment because it went up in the air instead of around and around betrayed a lack of imagination. Alvarez quit the leadership of his own LBL physics group in protest and got funding elsewhere, including from NASA, to see if HAPPE could fly. (It crashed.)

By the mid-1970s, Alvarez had his own Nobel Prize in Physics, and the lab had a new director. One day Alvarez was reading a magazine article about an experiment being done by an acquaintance of his. Stirling Colgate, toothpaste scion by birth and thermonuclear physicist by choice, had planted a 760-millimetre telescope on a surplus Nike missile turret in the New Mexico desert. The plan was to program it to automatically look at a different galaxy every three to ten seconds. The telescope would then transmit the information through a microwave link to the memory of an IBM computer on the campus of the New Mexico Institute of Mining and Technology, eighteen miles away. There, software would search the images for supernovae.

Automated astronomy wasn't entirely new. The era of the lone observer standing on a mountaintop, eye to eyepiece, staring into the abyss, was coming to an end. Since the invention of the telescope in the first decade of the seventeenth century, astronomers and telescope operators had guided their instruments by hand. Now astronomers were writing computer programs that manipulated the motions of the telescope — and did so with far greater sensitivity than the human hand, and eventually without the need for a human presence on site at all. The University of Wisconsin had an automated telescope. So did Michigan State University, and MIT too.

Supernova searches weren't new, either. They had been a staple of astronomy since Zwicky's initial observing programme in the 1930s. The 460-millimetre telescope on Mount Palomar that he helped design was still in use for that purpose, as was a nearby 1.2-metre, along with telescopes in Italy and Hungary.

What distinguished Colgate's project was the combination of the two ideas — automated telescope, supernova search. An astronomer looking for a supernova has to compare images of the same galaxy several weeks apart to see whether a dot of light has emerged. Traditional supernova searches required developing photographic plates by hand, then comparing them to previous exposures by eye, both time-consuming processes. In Colgate's system, a television-type sensor would replace photographic plates, and a computer would compare images almost instantly. In the end, his experiment didn't work. His hardware was fine; he had been able to make a telescope that could point where and when he wanted it to point. The problem was the software. The computer code required the equivalent of a hundred thousand FORTRAN statements, and Colgate was working more or less alone. Still, Alvarez saw, the potential was there. No: The *inevitability* was there — the era of the remote supernova survey.

Alvarez looked up from the magazine. "Stirling has been working on this project," he said, handing the article to a postdoc and former graduate advisee of his, Richard Muller. "And I think he's abandoning it. Talk to him. See what's going on."

By the time Perlmutter arrived at LBL in 1981 as a twenty-one-year-old graduate student, the stories of Muller's battles with the bureaucracy were legion. Year after year, LBL leadership would tell Muller that this was the last time he'd be getting funding, because if the US Department of Energy was going to cut anything from the LBL budget, it was going to be the speculative astrophysics project, and then that money would be gone from the LBL coffers for good. But Muller had read Jim Peebles's book *Physical Cosmology,* and he understood that the astronomy of the very big and the physics of the very small were becoming closer, even indistinguishable. It was the era of the cosmic microwave background. Of quasars, those mysterious sources of extremely high energy from the depths of the distant universe. Of pulsars, which provided evidence not only that Zwicky and Baade had been right all those decades earlier about the existence of neutron stars, but that these stars were spinning at a rate of hundreds of times a second. You couldn't study any of these phenomena without thinking about high-energy physics. Astronomy might not be the kind of high-energy physics that the lab had ever pursued, but it was quickly becoming the kind that the lab had always done. From the point of view of a Luis Alvarez or a Richard Muller, they weren't drifting towards a new discipline. The discipline was drifting towards them.

If anything, Perlmutter thought of himself as a born physicist—someone who wanted to know how the world worked on the most fundamental level, to discover the laws that united all of nature. Science courses had always been the easy part of his education; he almost hadn't needed to think about them. He studied science, enjoyed it . . . and then had plenty of time to indulge other interests. And what he thought on, when he wasn't thinking on the laws that united nature, were the "languages" that united humanity—literature, maths, music. So maybe he was a born philosopher, too. His parents were both professors—his father taught chemical engineering, his mother social work—who had elected to raise their family in an ethnically mixed neighborhood in Philadelphia, and he grew up listening to his parents and their friends talking about social issues and the latest books and films. He began practicing violin in the second grade. He joined the chorus. In secondary school he challenged himself to learn how to think like a writer—to learn the nature of narrative. And when he arrived at Harvard College in 1977, he assumed he would study both physics and philosophy.

The "physics stuff," he soon realized, was starting to get hard. University physics bore little relation to secondary school physics. He concluded that if he pursued physics and philosophy together, he would have no time for other courses, let alone a social life. He would have to make a choice.

If he chose philosophy, he couldn't do physics. But if he chose physics, he would still be doing philosophy, because in order to do physics you had to ask the big questions. Before science was science (the study of nature through close observation), it was philosophy (the study of nature through deep thought). Even as science had accumulated all manner of empirical scaffolding over the past few centuries, the guiding impulse of the scientist had remained constant: What is our relationship to the natural world? When Perlmutter joined Richard Muller's physics group at LBL in 1982 and had to choose among the eclectic programmes Muller was directing that were now acceptable to the institution — using planes to sample carbon in the atmosphere, measuring the gravitational deflection of starlight by Jupiter — he selected the nascent supernova survey because it seemed the kind of project that might lead to the biggest questions of all. Just as you were automatically doing philosophy if you were doing physics, you were automatically doing physics if you were doing astronomy. Instead of the nature of narrative, Perlmutter would be exploring the narrative of nature.

Alvarez had handed the idea of an automatic supernova survey to Muller on a whim, and now Muller handed that whim to his own postdoc, Carlton Pennypacker. By 1984, the Berkeley Automatic Supernova Search (BASS) team — Muller, Pennypacker, Perlmutter, and a few other graduate students — was operating on the 760-millimetre telescope at the university's Leuschner Observatory, in the hills a half-hour drive northeast of campus. Further technological advances that were particularly useful for supernova hunting were coming along all the time. When astronomers using photographic plates hunted for supernovae by eye, they used an optical device called a comparator. By rapidly switching back and forth between two images of a galaxy taken several weeks apart, the comparator would allow an astronomer to see whether any new pinpoint of light had appeared in the interim. The comparator *blinked* the two images. New computer technology, however, allowed astronomers to take all the light from the earlier image and remove it from the later one. It *subtracted* the first image from the second. If the computer signaled that a telltale bit of light remained, then a real live human being analysed the data. Sometimes the source of the bit of light was "local"—a fluctuation in the output, a cosmic ray from space striking the instrument, a subtraction error. Sometimes the source of the light was "astronomical"—asteroids, comets, variable stars. But once in a while the blip of light was a star erupting in a farewell explosion that stood out even against the background of all the tens or hundreds of billions of other stars in its host galaxy combined: that is, a supernova.

Or a Nemesis. In 1980, Luis Alvarez, along with his son Walter Alvarez, had hypothesized that the mass extinction of the dinosaurs 65 million years ago, at the cusp of the Cretaceous and Tertiary periods, had been caused by a comet or asteroid impact that had disrupted the global ecosystem. Then, in 1983, a pair of paleontologists announced that they had discovered evidence of a cycle of mass species extinctions every 26 million years. The following year, Muller and some colleagues published a paper speculating on the existence of a companion star to the Sun—Nemesis. Every 26 million years, they wrote, the highly elliptical orbit of Nemesis would bring it relatively near the Sun, and its gravitational influence would draw comets from the farthest reaches of the solar system into the orbital paths of the planets nearest the Sun, including Earth.

The idea wasn't as fanciful as it might seem; studies of Sun-like stars had shown that about 84 percent were in binary systems, meaning that the Sun, if solo, would be an anomaly. Muller assigned Perlmutter the task of looking for Nemesis (or the Death Star, as the media called it), and in 1986 Perlmutter completed his thesis, "An Astrometric Search for a Stellar Companion to the Sun." But the two projects shared a telescope and some other hardware, some of which Perlmutter helped design, as well as the search software, much of which he wrote, and when he was invited to stay at LBL as a postdoc, Perlmutter looked to supernovae for inspiration. A few months earlier, on May 17, 1986, BASS had bagged its first supernova, and that was good enough for him.

In 1981 the team had predicted a detection rate of one hundred supernovae per year, but scientific proposals were often overoptimistic. Besides, BASS supernovae were relatively nearby; they weren't going to be immediately useful for the big questions. Any changes in the universe's rate of expansion wouldn't be discernible unless astronomers could find standard candles in galaxies significantly farther than Hubble's sample, the deeper the better. How much those supernovae—and therefore their host galaxies—deviated from the straight-line Hubble diagram would tell astronomers the rate of deceleration. Thanks to BASS, Pennypacker and Perlmutter now knew they could do an automated supernova search; they had added two more supernovae in 1986 and another in 1987. But could they do an automated supernova search at cosmologically significant distances?

Muller himself thought such a project might be premature. But he was also a scientist who for years had entertained Luis Alvarez's imaginative flights and was willing to risk his own scientific reputation on a search for a Death Star. He gave his consent, and Pennypacker applied for funding for a camera he wanted to mount on a telescope in Australia. Or, rather, Pennypacker commissioned the camera, *then* applied for the funding. But the heavens opened over Berkeley and the NSF showered millions of dollars on the Center for Particle Astrophysics, and even though the names on the supernova proposal were Richard A. Muller and Carlton Pennypacker, the search for the fate of the universe was, from the start, Pennypacker's and Perlmutter's.

Supernovae remained attractive as potential standard candles for a couple of reasons. They're bright enough to be visible from the farthest recesses of space, meaning that astronomers can use them to probe deep into the history of the universe. And they operate within human time frames, their luminosity rising and falling over the course of weeks, meaning that, unlike most astronomical phenomena (such as the formation of a solar system or the coalescing of galaxies into a cluster), supernovae offer a soap opera that astronomers can actually watch.

But supernovae were also problematic for at least three reasons. As the LBL group put it, "they are rare, they are rapid, and they are random." In our own Milky Way galaxy, supernovae pop off at an average rate of maybe once a century. So astronomers pursuing supernovae have to devise a way to look at a great number of galaxies, whether individually in quick succession or in great gulps of the sky all at once — or, ideally, both: great gulps in quick succession. Supernovae also require a fast response. Once astronomers identify a supernova, they have to move quickly to do the necessary follow-up studies — not always possible when time on telescopes is assigned months in advance. And they're random. You never know where or when one's going to go off, so even if you could reserve follow-up time on telescopes months in advance, you wouldn't know whether you would have a supernova worth studying on that date.

Perlmutter and Pennypacker were already testing new subtraction software on the 1.5-metre telescope at Mount Palomar when they first heard that they weren't alone. The idea of chasing supernovae for clues about cosmology was half a century old, and now the widgets were out there to make that chase a reality, so they weren't entirely surprised that another group had been using distant supernovae to try to determine how much the expansion of the universe was changing over time.

Trying, and failing. From 1986 to 1988, three Danish astronomers, with the help of two British astronomers, took turns making a monthly trek to a 1.5-metre telescope at the European Southern Observatory in La Silla, Chile. They calculated that if they looked not at one galaxy at a time but at clusters of galaxies, they could beat the once-a-century-per-galaxy odds of finding the right kind of supernova. They selected clusters with well-established distances. And they timed their searches carefully, choosing the nights just before and after a new moon so that they were able not only to capitalize on dark skies but to compare images about twenty days apart, a period that, through happy coincidence, corresponds to the natural life (or, more aptly, death) cycle of the kind of supernova they wanted.

They found one. Possibly a second, on what would be their final night of observing, though they didn't bother to follow up that detection. They were ahead of their time, and, after two years, their time was up. Their telescope turned out to be too small for their purposes, their rate of discovery too low. Unless they could garner access to a larger telescope and a more powerful detector, they would need dozens of years to collect a suitable sample. Even at the rate of one good supernova a year, they would still need ten years, minimum, to complete their programme.

For the members of the Berkeley supernova team, the news of this failure raised a potentially fatal question: How could they reassure the review committees at the Center for Particle Astrophysics that their group could succeed where the Danish collaboration had failed?

First, the team stressed that the Danes *had* succeeded in finding a distant supernova—so distant that it broke the redshift record for a supernova, 0.31, meaning that it exploded about a quarter of the way back to the beginning of the universe (or 3.5 billion years ago). Second, the Berkeley team would be using better instrumentation. The Anglo-Australian Telescope in Siding Spring, Australia, had a 3.9-metre mirror, more than double the diameter—or four times the aperture, the light-collecting area—of the one the Danish team had used in Chile. And LBL was commissioning a much larger camera.

For insurance, Pennypacker invited Gerson Goldhaber onto the project. In 1933, when Goldhaber was nine, his family fled Germany. He lived in Cairo and then Jerusalem before emigrating to the United States to pursue a PhD at the University of Washington. Goldhaber had worked at LBL since 1953, playing key roles on the Bevatron and then, for the past twenty years, collaborating with the Stanford Linear Accelerator Center. He had made important discoveries He had guided teams that won the Nobel in Physics. As Pennypacker reasoned, "They would never shut down anything he did."

Problems began even before they could start observing. The contractor constructing the camera delivered a mirror that didn't fall within "tolerances," as opticians call the allowable imperfections. The second cut was spoiled when cleaning fluid spilled on the mirror. Finally, the third cut of the mirror worked.

Pennypacker, however, had ordered a camera without a filter, figuring that the more light he got, the better—and not understanding that if you want to compare the brightness of an object on different days, you need to observe in different filters in order to "equalize" the light level. After the crack technical crew at LBL designed an after-the-fact filter wheel, Pennypacker handed it to a graduate student, sent her to Australia, and provided her with the number to give to customs. When she got to customs, she went up to the two clerks and said, "I have this number." They looked at each other, then back at her. "Number for what?" one of them said. (A few days later the director of the observatory in Australia managed to convince the authorities to un-confiscate the wheel.)

Even when the team did perform supernova searches, the subsequent logistics were daunting. The on-site computers didn't have sufficient bandwidth for the Berkeley project's purposes, so team members had to take the computer data to the Sydney airport, fill out volumes of paperwork, and put the cargo on the next plane to San Francisco. There, someone else had to fill out more paperwork before claiming the package and driving it across the bay to Berkeley. Total travel time: forty-eight hours. Then the physicists at Berkeley needed another two days to search the images for supernova candidates, and then another day to study finding charts — maps that show all the known objects in a section of the sky — to see if they really were supernovae. Five days is a long time to wait if you want to schedule follow-up observations of an object that is fast disappearing. Before long, however, they managed to figure out a way to get their data to Berkeley without air travel: A team member at LBL would call up NASA, specify when the supernova search would need to be transmitting data from Australia, and ask if someone at NASA would please turn on the Internet then.

Not that the quality of the data or a delay in transmission mattered. In the course of two and a half years, Pennypacker and Perlmutter secured a dozen nights on the Anglo-Australian Telescope; of those, nine and a half were cloudy or had poor atmospheric conditions, or were needed for testing. And although they did manage to identify six candidates, the final count of actual supernovae was worse than the Danes': zero.

The Berkeley team had come up with an ambitious three-year plan, and they'd executed it. They had stretched the existing technology as far as they could under the circumstances, tweaking each widget until it had nothing left to give, and their efforts simply weren't enough.

Every few months the supernova search had to justify its existence as part of the Center for Particle Astrophysics to an internal Program Advisory Committee. Every few months it also had to justify its existence to an External Advisory Board. That justification now took the form of the team's ability to secure time on the 2.5-metre Isaac Newton Telescope, in the Canary Islands, off the northwest coast of Africa. The telescope was slightly smaller than the one in Siding Spring, but the camera would be bigger, and the weather would be better. Muller himself made the pitch to Bernard Sadoulet, the director of the centre: "Look, two or three years from now, we will be delivering supernovas. We will be making real measurements. We will have results. I guarantee that. By the time the initial funding for the Center for Particle Astrophysics runs out, we will have something real to show. And you must understand that. *You must know that.*"

The supernova search got its reprieve. If you were a veteran of the project, you could consider yourself on probation yet one more time. Newcomers, however, wondered what they'd gotten themselves into. Group meetings were held in an office where there weren't enough chairs. You might find yourself sitting at a computer, quietly typing away, when a higher-up would tell you that the project had exceeded its computer allotment and to either shut down the computer *now* or someone would be along shortly to pull the plug. One graduate student read the recruitment brochure, liked the idea of "weighing the universe," and committed to the project—only to learn that the search had yet to produce one supernova. A postdoc arrived for his first day on the job to find a note on his desk from Perlmutter, saying that he'd gone to the Canary Islands and asking the postdoc to use a finding chart to choose fields for Perlmutter to target. The postdoc stared at the note. He had trained as a particle physicist; he didn't know what a finding chart was.

And then Pennypacker—his words—"blew up the budget." He had developed a habit of spending money he hoped would materialize in the next round of budgets. This time, however, he misread a ledger, spent money he didn't have, and then spent it again.

By his own admission, Pennypacker wasn't leadership material, at least not of the kind required to run an unorthodox and high-risk project. People loved collaborating with him; he was enthusiastic and affable and smart and visionary. The ability to translate those virtues into the words that a review committee or an administrator wanted to hear, however, eluded him. So did a fundamental understanding of administrative details. If the supernova project were to continue, he was made to understand, it would have to do so under different leadership.

Robert Cahn, the new director of the Physics Division at LBL, first approached the senior researcher on the project, Gerson Goldhaber. But for Goldhaber the chance to work on supernovae had represented a freedom from the kind of responsibilities he'd held for four decades at behemoth particle accelerators. Muller had moved on. The next choice was Saul Perlmutter. Cahn consulted with Muller: Was the kid ready? Muller thought perhaps so. Twice, Muller said, he'd had the experience of going to Perlmutter with what he thought was a conceptual breakthrough, and Saul had said, "Ah! That's a very interesting idea," then pulled out a notebook and flipped to the page where he'd already seen the idea through and found that it didn't work.

In March 1992, Perlmutter went to the Isaac Newton Telescope to take the first set of images. In late April and early May, Pennypacker went to the INT to make follow-up observations of the same fields while Perlmutter stayed in Berkeley and waited for data to arrive via the BITNET. As the sun was setting over the Atlantic — midafternoon in Berkeley — Perlmutter and a couple of team members would settle into the seats before the high-quality image display in the deliberately overcooled computer centre in the basement at LBL, bundled in jumpers and jackets, and start to sort through the software. By ten or eleven in the evening, Perlmutter was alone and the images would begin to emerge on his screen. Each image held hundreds of galaxies; by the end of the night he would collect dozens of images. He printed out each one, just in case. Sometimes the computer told him that a blip of light had appeared that hadn't been there the previous month, and he would bend close to the screen and try to figure out what was wrong. The view of the wide-field camera distorted the geometry, so he didn't trust blips near the edge of the frame. Sometimes a blip would be too near the centre of a galaxy, meaning that its light would be impossible to distinguish from the background light during follow-up observations. Sometimes the blip turned out to be an asteroid. One night he found a blip that he couldn't discount, and he had to wonder what obvious error he was overlooking, but he couldn't think of one, so he asked himself what *subtle* error he was overlooking, but he couldn't think of one, and then he wondered what he was doing wrong, when suddenly he realized, "Wait—this is what we're supposed to be looking for": the potential supernova you can't throw away.

For corroboration he had to wait until his collaborators showed up in the morning, and even then it wasn't a moment for celebration, partly because they couldn't be absolutely sure, and partly because the work had just begun. The data was worthless without follow-up observations that would tell them whether it was indeed a supernova, and if so, at what redshift.

Some of those observations they could do on their own, in the days that followed, while they still had time on the INT. Others required them to find out which astronomers were observing on the big telescopes around the world, figure out whether anyone in the LBL operation might be friends with them, then phone them in the middle of an observing run they had been planning for six months or a year, to plead with them to drop everything and point their telescope somewhere else. In this regard, Perlmutter was singularly talented. Nobody worked the middle-of-the-night phone calls to astronomers on other continents like he did. He was persistent, and he was persuasive, and he was impervious to rejection or insult. He literally wouldn't take no for an answer. Sometimes the plea elicited a laugh, sometimes an outburst of anger. But sometimes the plea elicited data—just enough to tell them that the blip was still there, and fading. They had a supernova.

Still no cause for celebration. Again, the data was meaningless for cosmology unless they knew how distant the supernova was—its redshift. For that, they would need a spectroscopic analysis. Twelve times, at four observatories around the world, astronomers agreed to make the follow-up observations. Eleven times the weather didn't cooperate. The twelfth, the instrument malfunctioned.

As spring stretched into summer, Pennypacker began to think of his team as characters in *The Treasure of the Sierra Madre:* fortune hunters wandering the desert in search of gold. And they find it—the prospectors discover their vein; the astronomers detect their supernova. And then the gold dust slips through their fingers and blows away in the wind. Walter Huston or Humphrey Bogart or Tim Holt says *Thanks anyway* to a pal in a faraway observatory, then slowly hangs up the phone.

One night in late August, Pennypacker and Perlmutter called Richard Ellis, a friend to the team as well as a British veteran of the Danish observations in the late 1980s. Ellis snapped at them. Didn't they know that observing conditions in the Canary Islands had been bad lately, and that he and the other observers were already inundated with requests for make-up observations from astronomers who had actually had time on the telescope—unlike the Berkeley team?

Then he went and made the observation. On 29 August 1992, Ellis took out his finding chart and, working on the 4.2-metre William Herschel Telescope, he and a postdoc took two half-hour spectra. When they were done, Ellis got Pennypacker on the phone in Berkeley.

The old record redshift, the one set by the Danish team, had been 0.31, corresponding to roughly 3.5 billion years ago. The new record redshift was 0.458, or 4.7 billion years ago.

Pennypacker let out a whoop. Six years after he and Perlmutter had discussed collaborating on a search for cosmologically significant supernovae, they hadn't found the weight or the shape or the fate of the universe.

But they were in the game.

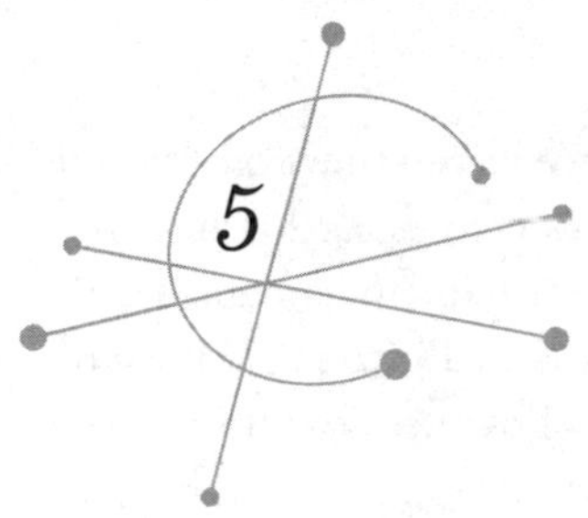


# Staying in the Game

IN EARLY 1994, a couple of astronomers got to talking. Brian Schmidt had just completed a PhD thesis on supernovae at Harvard's Center for Astrophysics, and he was thinking about ideas for his next project as a postdoc. Nicholas Suntzeff had been an astronomer at the Cerro Tololo Inter-American Observatory in Chile since 1986, and he had been working on a supernova survey since 1989. As supernova specialists they had both been following the efforts of Berkeley's supernova project. Now, as they sat in the air-conditioned computer room at the observatory headquarters in the Chilean coastal town of La Serena, Schmidt mentioned that he'd been thinking about putting together a team to go up against LBL's.

Suntzeff didn't hesitate: "Can I be part of that?"

Now *that,* Schmidt thought, is the mark of a good problem in science. It's not when people say, "Oh, that's interesting." It's when they say, "*Ooo,* can I be part of that?"

Schmidt had to give Saul Perlmutter and the Berkeley team credit. They had seen that, thanks to advances in technology, supernovae might finally be used to do cosmology, and they were succeeding against enormous odds. They had been in the right place at the right time. But were they the right team?

Like many other astronomers, Schmidt had been sceptical that physicists — even physicists-turned-astrophysicists — would be able to consistently find distant supernovae. But even after the LBL team had found its first supernova, Schmidt and other astronomers remained sceptical that physicists-turned-astrophysicists — no matter how brilliant—could perform the kinds of follow-up observations and analyses that routinely strained even their own hard-won expertise. Seemingly everybody in the supernova game had been on the receiving end of a middle-of-the-night phone call from Saul, asking them to drop everything and perform a follow-up observation of a supernova candidate. Perlmutter had gotten a reputation in the community for being preternaturally persistent. But in Suntzeff's experience, every time he slewed his telescope to Perlmutter's target, the field was empty. "Must be too faint," Suntzeff would say diplomatically.

Schmidt and Suntzeff grabbed the nearest blue-and-gray sheet of IBM computer printout, flipped it over, and began scribbling. They continued the conversation in Suntzeff's office later that day and the next day as well, laying out their plan of attack.

Suntzeff, they decided, would be in charge of the observing. He would find the supernova candidates and do the follow-up measurements. Schmidt would be in charge of the analysis. He would take some existing software and create a new code that would clean the images, do the subtraction, and isolate the supernovae.

Suntzeff turned to Schmidt. "How long will it take you to write the new code?"

Schmidt could carry himself as a cocky young astronomer, but his wiseacre's side-of-the-mouth way of talking suggested not so much arrogance as irony. Suntzeff preferred to think of him as constitutionally optimistic. Yet even Schmidt had to hesitate. Then he reminded himself: *Saul's doing it.*

"Two months," he answered.

On his return to Harvard, Schmidt disappeared into his office for hours at a time, day after day, week after week, writing the code. But he also circulated through the halls, stopping colleagues and dropping into offices, letting a select group know that he and Nick Suntzeff were putting together a team to catch Saul. In each case he got the same response, expressed with the same level of eagerness: *Can I be part of that?*

Robert Kirshner wouldn't even have to ask. He had been a student of supernovae since 1970, longer than some of his students had been alive. At forty-four, he was now an elder statesman in astronomy, the chairman of the Astronomy Department at Harvard. He had long experience getting money out of the National Science Foundation, reserving time on the world's best telescopes, and helping to set policy for the Space Telescope Science Institute, the science and operations centre for the Hubble Space Telescope. He was one of the world's foremost supernova experts, as well as the mentor to several generations of supernova experts — the graduate students he had recruited and the postdocs he had hired for his private duchy within the Harvard-Smithsonian Center for Astrophysics, half a mile up Garden Street from Harvard Square. When *Nature* received the Danish group's preliminary results in 1988, it was Kirshner whom the journal asked to privately review the paper and then publicly write an accompanying news analysis. When Berkeley's Center for Particle Astrophysics convened an External Advisory Board and needed a supernova guru, it was Kirshner who got the call. When Perlmutter et al. submitted a paper to the *Astrophysical Journal Letters* analysing their 1992 supernova, it was Kirshner whom the editors asked to serve as referee.

To all his peer evaluations Kirshner brought a deep scepticism, born of his own decades of experience, about the ability of anyone to perform the near-surgical task of supernova analysis. Kirshner could be amusing; in casual conversation he often made exaggerated facial expressions, adopted accents, whinnied at his own jokes. His presentations at conferences were reliably witty and well-attended. But when it came to supernovae, and to what you needed to know to do supernova astronomy, Kirshner could be exacting, even bruising.

But he had a point — several points, actually. If you wanted to do supernovae, you had to know spectroscopy — the analysis of an astronomical object's spectrum of light that identified its chemical composition as well as its motion towards or away from you. You had to know photometry — the tedious, difficult determination of an object's brightness. You had to account for dust, either within the supernova's host galaxy or somewhere along the line of sight between the supernova and the observer. Sometimes dust was there, sometimes not. If it was there, it would dim or redden the light from the supernova. And if you didn't know the extent to which dust was polluting the light, you wouldn't know how much to trust your data.

For the Berkeley supernova group, however, Kirshner reserved a special level of scepticism. As far as he was concerned, they were doing poor work that was giving his area of expertise a bad name.

From the start, Kirshner had his doubts about a bunch of particle physicists trying to do astronomy, adopting it as if it were a hobby rather than a science you needed to perfect over a lifetime. So far he'd seen nothing to ease those concerns. In the 1980s, Richard Muller had diverted time from the supernova survey at the Leuschner Observatory to pursue his Nemesis project. The discovery of a companion star to the Sun, if he made it, would be momentous, but it was so unlikely that the effort seemed almost a capricious use of precious telescope time. In 1989, Muller, Pennypacker, and Perlmutter got the attention of astronomers around the world by concluding that the famous supernova 1987A—the first naked-eye supernova in four hundred years—had left behind a pulsar, a neutron star spinning hundreds of times per second. The "evidence" turned out to be an instrument error. And then came the embarrassment that Kirshner got to witness for himself, as a member of the External Advisory Board: a three-year attempt to find distant supernovae at the Anglo-Australian Telescope that had come up empty.

Hundreds of thousands of dollars: money enough to fund dozens of more modest and more practical astronomical projects: empty.

It wouldn't be entirely fair to say that particle physics operates according to the principle *Get funding first, ask questions later.* But it wouldn't quite be inaccurate, either. Projects in particle physics routinely involve dozens, hundreds, even thousands of participants, and require machines that manufacture ultra-energetic pyrotechnics that the universe hasn't seen since its megacompact first fraction of a second of existence. The Berkeley supernova search wasn't operating on that scale, but other projects at LBL were, and the lab itself had long been the world's foremost proponent of that work ethic. Particle physicists can somewhat afford to bulldoze ahead, confident that between their million-ton hardware and their collective brainpower they'll find the answer to any question they might ask. And the first question that the LBL team had asked was: Can we find distant supernovae?

It was, Kirshner thought, the wrong question to ask first. The right one was whether distant supernovae were worth finding. Could they really serve as standard candles?

The recent history of astronomy held a couple of cautionary tales for the standard-candle-bearer. Having discovered evidence for the expansion of the universe, Edwin Hubble spent much of the last twenty years of his life working under the assumption that galaxies might be standard candles, even though they weren't entirely uniform. Maybe they were similar enough that he could use them to discern the universe's shape and fate. Walter Baade, one of his Mount Wilson colleagues as well as Fritz Zwicky's collaborator on the 1934 "super-nova" paper, argued that Hubble had it backwards: "You must understand the galaxies before you can get the geometry right." Allan Sandage, Hubble's protégé and, upon his death in 1953, his successor at the Mount Wilson and Palomar Observatories, would later write, "Hubble clearly understood this, but rather than be stopped because this part of his subject was 30 years before its time, he pushed ahead with an abandonment known to pioneers in any milieu who try to reach Everest without proper equipment."

Then it was Sandage's turn. For a quarter of a century, he and the Swiss astronomer Gustav A. Tammann pursued an alternate candidate for standard candles. If galaxies themselves weren't uniform enough, then maybe clusters of galaxies were—or, more precisely, the brightest galaxy within each cluster. But this proposition, too, suffered from an insufficient understanding of galaxy mechanics. Some galaxies would grow dimmer with age, as their stars died out, while other galaxies would grow brighter with age, as they merged with smaller galaxies. Unable to reliably tell the difference, and wary of other factors they couldn't begin to guess, Sandage and Tammann turned back from the summit. "Essentially," Sandage announced to his colleagues in cosmology in 1984, at a conference on the expansion rate of the universe, "we have failed."

Kirshner never passed up an opportunity to point out that the same fundamental lack of understanding of underlying processes could easily sabotage the usefulness of supernovae for cosmic measurements. Already astronomers had determined that supernovae belong to two classes—and possibly more.

One class was the kind that Zwicky and Baade had prophesied—one that results in the birth of a neutron star. It was the kind Zwicky assumed he was finding in his 1930s survey of "star suicides." In 1940, however, Rudolph Minkowski at Mount Wilson took a spectroscope of a supernova that was different from the spectroscopic analyses of Zwicky's supernovae. Minkowski's supernova showed the presence of hydrogen. Zwicky's supernovae did not. They were clearly different types of supernovae.

Since then astronomers had come to think that one type of supernova—the type that Zwicky and Baade had predicted in 1934, that Zwicky thought he was observing in 1936 and 1937, and that Minkowski did observe in 1940—was the result of a chain of nuclear processes in a star several times the mass of the Sun, leading to a 40,000-miles-per-second implosion.

The other type—the type that Zwicky observed—begins life as a hydrogen-rich star like our own Sun. As it ages, the Sun will shed its outer hydrogen layer while its core contracts under gravitational pressure. In the end, only the core will remain—a shrunken skull called a white dwarf, with the mass of the Sun packed into the volume of Earth. If a white dwarf had a companion star (and most stars in our galaxy do), then at this point it might start to siphon gas off the other star. In the 1930s, the Indian mathematician Subrahmanyan Chandrasekhar calculated that when a star of this kind reaches a certain size—1.4 times the mass of the Sun, or the Chandrasekhar limit—it will begin to collapse of its own weight. The gravitational pressure will destabilize its chemical composition, leading to a thermonuclear explosion.

Through a telescope on Earth, the two types would look the same, even though one is an implosion and the other is an explosion. But a spectroscope would show the difference—hydrogen or no hydrogen, Type II or Type I. For astronomers, the uniformity of Type I supernovae offered the possibility that this type might be a standard candle. Since these supernovae all began as a single kind of star, a white dwarf, that had reached a uniform mass, the Chandrasekhar limit, maybe their explosions had the same luminosity.

In the 1980s, however, the clear distinction between Type I and Type II began to blur. Spectroscopic analysis of three supernovae—one each in 1983, 1984, and 1985—showed that they consisted of huge amounts of calcium and oxygen, consistent with the interiors of massive stars that end their lives as Type II supernovae, but no hydrogen, consistent with white dwarfs that end their lives as Type I supernovae. Some astronomers, including Kirshner, suggested that they were seeing a third type of supernova, essentially a hybrid of the other two. It was the product of a core collapse that had already lost its outer shell: a hydrogen-free implosion.

They added this specimen to the Type I column, calling it Type Ib. The old Type I, a thermonuclear explosion with no hydrogen, was now Type Ia.

In 1991, even *that* classification—Type Ia—began to blur. On 13 April, five amateur observers in four locations around the world discovered a supernova designated 1991T.* On 9 December, an amateur astronomer in Japan discovered a supernova designated 1991bg. Follow-up spectroscopic observations by professional astronomers—including Kirshner, on 16 April, for 1991T—showed that they were both Type Ia supernovae. But their luminosities differed widely. Supernova 1991T was much brighter than the usual Type Ia at its particular distance, and 1991bg was much dimmer than the usual Type Ia at *its* particular distance. Astronomers could rule out the possibility that they were simply miscalculating distances: The dimmer supernova was ten times dimmer than a supernova observed in 1957 in *the same galaxy*.

Astronomers began to suspect that while each supernova in the universe might be a Type Ia, Type Ib, or Type II, the types themselves might be more like families. The supernovae within a family share traits, but they're not identical; they're more like siblings than clones. For astronomers hoping to adopt Type Ia supernovae as standard candles, Kirshner wrote, the problem "was serious and real." You couldn't ignore it.

And the Berkeley group didn't ignore it. In her 1992 PhD thesis a team member summarized the collaboration's general attitude towards the problem: "There is still some contention" about

* Supernovae receive alphabetical labels based on the order of discovery within a year, first uppercase once through the alphabet (A, B, C . . . X, Y, Z), then back to the beginning of the alphabet but lowercase and doubling up (aa . . . az, ba . . . bz).

"whether individual SNe Ia do not fit the model," but, she added, echoing the chorus that Kirshner had heard from the LBL group again and again, "it is clear that the overwhelming majority of SNe Ia are strikingly similar."

Clear? Not to Kirshner, and he was the expert—a "realist," as he liked to call himself, not a wishful thinker.

In his role as a member of the External Advisory Board of the Center for Particle Astrophysics since the late 1980s, Kirshner emphasized that the Berkeley search team hadn't yet found a supernova, needed to be careful about photometry, couldn't account for dust—and didn't know whether Type Ia supernovae were standard candles.

Then in 1992 the LBL group found their first supernova. In his referee's report for *Astrophysical Journal Letters,* Kirshner complained that they still needed to be careful about photometry, still couldn't account for dust—and still didn't know whether Type Ia supernovae were standard candles. All they had shown, he thought, was that one could find supernovae distant enough that one could, *in principle,* do cosmology with them. But the Danes had done that, too, and they'd done it four years earlier. What the LBL team hadn't shown, in Kirshner's reading of the paper, was that one could find supernovae distant enough that one could *in fact* do cosmology with them.

He sent the paper back for a simple reason: "They hadn't yet learned anything about cosmology"—basically, that you couldn't assume exploding white dwarfs were perfect standard candles. They *weren't* perfect standard candles. The best you could hope was that somebody, someday, would figure out whether Type Ia supernovae, however imperfect, might be just good enough.

In secondary school in Marin County in the late 1960s, Boris Nicholaevich Suntzeff Evdokimoff played on the same varsity soccer team as his good friend Robin Williams. At Stanford in the 1970s, he regularly competed on the tennis court with—and lost to—future astronaut Sally Ride. What was really cool, though, was that as a Carnegie Fellow in the early 1980s he got to talk astronomy with Allan Sandage.

Suntzeff loved historical connections in astronomy. A great-uncle of his had gone to school in Russia with Otto Struve, the descendant of a line of prominent astronomers. Struve fled Russia and the Bolsheviks at the time of the revolution and wound up in Turkey, impoverished, until a relative put him in touch with the director of the Yerkes Observatory in Wisconsin, who offered him a job as a spectroscopist. Struve later became director of the observatory, as well as McDonald Observatory in Texas and Leuschner Observatory in Berkeley. Suntzeff's family also fled Russia, though they headed in the other direction, to China and, eventually, San Francisco. There Suntzeff's grandmother reunited with Otto Struve. Small world.

And now Nick Suntzeff would be doing his part to make astronomy a bit more intimate. He had applied for a Carnegie Fellowship for just that reason: to spend time with Sandage at the headquarters of the Carnegie Institution's Mount Wilson and Palomar Observatories. There, on Santa Barbara Street in Pasadena, Edwin Hubble had figured out, in 1923, that the Milky Way was just one among a multitude of galaxies in the universe, and then, in 1929, that the universe was expanding. Allan Sandage arrived there in 1948, aged twenty-two, as a graduate student at Caltech. Over the next four years Sandage advanced from apprentice to assistant to Hubble's heir.

"There are only two numbers to measure in cosmology!" Sandage often said to Suntzeff, evoking the title of an influential article he'd written for *Physics Today* in 1970, "Cosmology: The Search for Two Numbers." The first number was the Hubble constant. The 45-degree straight line that Hubble plotted for the distances of galaxies and their redshifts — the farther the galaxy, the greater its velocity receding from us — implied a relationship you could quantify. If you knew how distant a galaxy was, then you should be able to know how much faster it would appear to be receding, and vice versa.

In the 1930s, Hubble himself estimated that galaxies were receding at a rate that was increasing 500 kilometres for every megaparsec (a unit of length in astronomy equal to 3.262 million light-years). That rate, unfortunately, corresponded to a universe that would be about two billion years old — which would make the universe younger than the three billion years that geologists had pegged as the age of the Earth. This disparity did nothing to help cosmology's reputation as a nascent science. But Hubble himself regarded his observations only as a "preliminary reconnaissance"; to do cosmology properly, he would have to keep seeking nebulae as far as the 2.5-metre telescope on Mount Wilson would allow, and then, eventually, as far as the 5.1-metre telescope on Mount Palomar, outside San Diego, would allow.

The 5.1-metre Hale Telescope saw first light in 1948, which happened to be the same year Sandage arrived at the Carnegie Observatories. But Sandage's timing was fortuitous in another way as well. The "monks and priests," as he called the first generation of Carnegie astronomers, were ready to retire. Up there, at the observatory on Mount Wilson, Sandage could dwell among his gods, astronomers who knew they'd "arrived," as Sandage would say, when they found their napkin not clipped to a clothespin but tucked inside a wooden ring inscribed with their name. And down here, on Santa Barbara Street, Sandage could inspect for himself "the plates of Moses"—the vast archives of photographic records, and an apt metaphor in more ways than one. Like Moses, Edwin Hubble had come down from the mountain bearing new laws of nature. But also like Moses, Hubble had to wander the desert for decades, only to die within sight of the Promised Land.

Hubble suffered a major heart attack in the summer of 1949, aged sixty, just six months after making his initial observations with the 5.1-metre telescope. To his assistant fell the responsibility for executing one of the most ambitious scientific programmes in history. Sandage found that the distances to the nearest galaxies were greater than Hubble had calculated, a correction that in turn affected Sandage's interpretation of more distant galaxies, which in turn affected his interpretation of even more distant galaxies. Distance dominoes fell as far as the 5.1-metre Palomar telescope could see. After Hubble's death in 1953, Sandage and his collaborators derived a Hubble constant of 180—a value he continued to revise downwards over the decades, until, by the time Suntzeff arrived at the Carnegie Observatories in the early 1980s, he had satisfied himself that the Hubble constant was around 50 to 55.

Despite its name, the Hubble constant wasn't a constant—a value unchanging over time. It told you only how fast the universe was expanding now—its *current* rate of expansion—and for this reason astronomers sometimes referred to it as the Hubble parameter. It told you nothing, however, about how much the expansion rate was changing over time. That value—Sandage's second number—astronomers called the deceleration parameter because it would tell you to what extent the universe was slowing down. From the Hubble parameter you could extrapolate backwards into the past and, depending on the amount of matter in the universe, derive the universe's age. From the deceleration parameter you could extrapolate forward into the future and, depending on the amount of matter in the universe, derive the universe's fate. In that sense, there *were* only two numbers to measure in cosmology: the alpha and the omega of the universe.

Both measurements would require a standard candle, and at the time that Suntzeff received his Carnegie Fellowship in 1982, Sandage (along with Gustav Tammann) had settled on supernovae. Sometimes Suntzeff and Sandage would be in Chile at the same time, at the Carnegie's Las Campanas facility, Sandage working on one telescope, Suntzeff another, and Sandage would ask Suntzeff to check whether a speck on a photographic plate was really a supernova. A dozen times Suntzeff swung the telescope to perform follow-up observations, and eleven times he had to break the news to Sandage: no supernova. In the end, Sandage figured out that he literally lacked the "proper equipment." The photographic plates were flawed. When he couldn't get Kodak to meet his exacting specifications for supernova searches, he abandoned the project.

But by then Suntzeff himself had become intrigued by supernovae. On cloudy nights at the observatory he would retire to the library and catch up on the supernova literature, or seek advice from Uncle Allan, as everybody called Sandage. The time was coming for Sandage to pass down to the next generation the programme that Hubble had passed down to him. He had lost some sight in his right eye, the one he had pressed to an eyepiece for four decades, and his sense of balance was off, a hazard on an observing platform metres above a concrete floor. Soon he would have to pack up his eyepiece, pocket his napkin ring, and come down from the mountain.

Besides, Sandage could see that his way of doing astronomy was coming to an end. For the first two centuries after the invention of the telescope, astronomers had to rely on nothing other than the light that hit their eyes at any one moment, and then that light was gone. Astronomers could draw what they had seen. They could capture it in words. They could record measurements to designate the location of an object or describe its motions. But what they saw—the light itself, the visual representation of the object in a moment in time—was gone.

The invention of photography in the mid-1800s radically changed that relationship between observers and their observations. For astronomy, photographs had an obvious advantage over the eye. A photograph preserved what an astronomer saw. It preserved the light itself, and therefore the image of the object at one particular moment. Astronomers could refer back not only to what they had drawn or captured in words or recorded as maths, but to what they had actually seen. And then so could any other astronomer, now or in the future.

But photography didn't just allow astronomers to collect light. It allowed them to collect light *over time.* Light didn't just land on the photographic plate; it landed and stayed there, and then more light landed and stayed there, and then more light. The sources of light were so faint your eyes couldn't see them, even with the help of a telescope, but the photographic plate could, because it was acting not like a moment-to-moment sensor but like a sponge. It could soak up light all night long. The longer the exposure, the greater the amount of light on the plate; the greater the amount of light, the deeper the view.

But now the charge-coupled device, or CCD, promised to do for the photographic plate what the photographic plate had done for the eyeball. A CCD consists of a small wafer of silicon that collects light digitally; one photon creates one electrical charge. A photographic plate is sensitive to 1 or 2 percent of the available photons; a CCD can approach 100 percent. For any aspect of astronomy, the advantages were obvious. Digital technology meant that you could process the images with computers, and more light meant that you could see farther and collect data faster. But for supernova searches, as Sandage explained to Suntzeff, the CCD came with a bonus.

The usefulness of a supernova for cosmology depends in large part on its light curve, a graph that shows the rise and fall of the luminosity of a supernova over time. Every supernova light curve rises abruptly over a matter of days as the supernova climbs toward maximum luminosity, then falls gradually as the supernova fades. But because each type of supernova releases its own distinctive cocktail of elements (hydrogen or no hydrogen, for example) and emerges out of a specific process (explosion or implosion), its light rises and falls in a signature pattern. To trace that pattern, you want to know when the curve peaks—when the brightness reaches maximum—so you need to be fortunate enough to discover the supernova on its way up. To chart the curve, you then need to make multiple follow-up observations—the more observations, the more data points you can plot on the graph; the more points, the more reliable the curve. But those observations are reliable only insofar as you can be sure how bright the light from the supernova is, and the accuracy of that measurement depends on how well you can distinguish the supernova light from the light of the host galaxy. A technology that allows you to make more observations and then quantifies those observations pixel by pixel can go a long way towards reducing the margin of error. The speed and precision of CCD technology, Sandage said, were going to make a light curve into a graceful, unambiguous arc—to the eye of a photometrist like Suntzeff, a work of art.

Suntzeff was already familiar with CCD technology. When he completed his Carnegie Fellowship in 1986, he became a staff astronomer at the Cerro Tololo Inter-American Observatory (a division of the U.S. National Optical Astronomy Observatory) in Chile; he was recruited by Mark Phillips, a good friend from his studies—they'd both been at the University of California, Santa Cruz, in the 1970s. Suntzeff's first mission was to install a CCD on a telescope, and he teamed up with Phillips to test the equipment on supernova 1986G. Suntzeff would do the observing and photometry, and Phillips would do the comparisons with light curves from other supernovae.

Suntzeff expected the result to be historic. As far as he and Phillips knew, theirs was the first "modern" light curve, meaning the first one obtained with a CCD. Historic though it was, the result was disappointing. The light curve for 1986G seemed to be significantly different from other Type Ia light curves. The supernova appeared to be fainter than it should be at its redshift, and the light curve looked as if it rose and fell more steeply than other Type Ia curves.

Part of the problem with being a scientific pioneer is that you have a compromised historical sample. The only light-curve comparisons that Phillips and Suntzeff could make had to come from photographic plates. They didn't know whether their odd CCD light curve said more about Type Ia supernovae or about CCD technology. Still, the two astronomers were confident enough in their corrections to the data that they concluded, in a paper they published the following year, that Type Ia supernovae probably varied too much in luminosity to serve as standard candles.

But as frustrating as the result was, Phillips and Suntzeff also sensed an opportunity. Their job would be to convince the community that Type Ia supernovae weren't standard candles—or to convince themselves that they had been wrong, and that Type Ia *were* standard candles after all. Either way, the two astronomers were in the supernova game now.

Their timing couldn't have been better. On 23 February of the following year, 1987, a supernova went off right overhead. SN 1987A appeared in the Large Magellanic Cloud, one of the few galaxies visible to the unaided eye—and only from the Southern Hemisphere. It was the first unaided-eye supernova since 1604, and among astronomers it prompted a worldwide viewing party. It wasn't a Type Ia, the explosive kind of supernova that Phillips and Suntzeff had studied. It was a Type II, the implosive kind. Still, on the basis of their access to a CCD in the Southern Hemisphere and their co-authorship on the SN 1986G paper, they found themselves assuming the role of what they facetiously called "the local supernova experts."

In July 1989, they attended a two-week supernova workshop at their alma mater, UC Santa Cruz. The topic for the first week was 1987A, but since the workshop was sure to attract just about every supernova expert in the world—all fifty of them—the organizers added a second week on supernova topics other than 1987A. By this point, virtually everyone at the conference had been working on 1987A nonstop for more than two years. They had results: observations, interpretations, theories. But they also had core-collapse-supernova fatigue. What about explosive supernovae? What about

Type Ia? The first week of the meeting would be for work; the second week, fun.

Sometimes at conferences the most productive work happens in the hallway between sessions, or over a beer in the evening. For Suntzeff, it happened during a conversation with an old friend. "There are only two numbers to measure in cosmology!" Uncle Allan boomed at him, and while Suntzeff didn't think much of the comment at the time, he recalled it later, back home in La Serena, when a junior staff member at the observatory mentioned an idea for a project.

Mario Hamuy had arrived at Cerro Tololo as a research assistant on 27 February 1987—three days after 1987A blossomed in the Large Magellanic Cloud. The original, pre-1987A plan for Hamuy was that, as the new hire, he would go to the mountain and spend a few days acclimating himself to the instruments. Instead, the director of the observatory sent him to the mountain to observe 1987A and only 1987A. By the time he returned to La Serena a month later, Hamuy was, if not yet a bona fide supernova expert, at least a supernova enthusiast above and beyond his characteristic enthusiasm.

Now he explained to Suntzeff and Phillips that he had attended a talk at Santa Cruz by Bruno Leibundgut, a Swiss astronomer with a fresh PhD. Suntzeff and Phillips knew Leibundgut from observing runs in Chile in the early and mid-1980s, when he was a graduate student working with Gustav Tammann. They had attended his talk, too, and Leibundgut told them afterwards that he'd been a late addition to the schedule. Bob Kirshner had recently hired him as a postdoc at Harvard, starting that fall, and at some point during the meeting Kirshner had turned in his seat and casually asked Leibundgut when he was giving a talk. Leibundgut answered that he wasn't; Kirshner told him that he was, now: He could have Kirshner's slot. And so Leibundgut wound up telling the world's supernova experts about his doctoral thesis: a template for Type Ia supernovae suggesting that they might be standard candles after all.

For Phillips and Suntzeff, the talk was part of a long-term, ongoing conversation in the supernova community. They had made their own contribution through their work on 1986G. For Hamuy, however, the talk provided a vision for the future. Listening to Leibundgut, he recalled that his graduate advisor at the University of Chile, José Maza, had coordinated a supernova survey in the late 1970s and early 1980s. Maybe the time had come to revive the idea of a supernova survey from the Southern Hemisphere, this time using the superior CCD technology. On his return from Santa Cruz, he had approached Maza with the idea, and Maza agreed to help. Now Hamuy wanted to know what Phillips and Suntzeff thought.

Phillips told him he thought it might be a good idea, but, he cautioned, a supernova survey from the Southern Hemisphere had to be something more than a supernova survey from the Southern Hemisphere. At which point Suntzeff thought, "There are only two numbers to measure in cosmology."

Maybe they were wrong in thinking that Type Ia supernovae were not standard candles. Maybe Leibundgut, who after all had been studying other Type Ia while they were busy with the Type II 1987A, was right. And if he was right, then maybe they could use nearby Type Ia supernovae to measure the Hubble parameter—the current rate of the universe's expansion. And if that programme worked, they could go to farther supernovae to measure the deceleration parameter—the rate at which the expansion was slowing down.

Hamuy devised the logistics. The survey would be a collaboration between two observatories, Cerro Calán, the university's observatory, in Santiago, in the middle of the city, and Cerro Tololo, his current employer—hence the Calán/Tololo survey. Ideally, a supernova search would combine the widest-field camera with the latest CCD technology, but that option wasn't available to the collaboration. Instead they had to choose between a telescope that couldn't accommodate a CCD camera but had a wide-field view and a telescope that could accommodate a CCD but had a narrow-field view. They chose the wide-field, no-CCD view, the 610-millimetre Curtis Schmidt Telescope on Cerro Tololo. When hunting prey as rare and elusive as supernovae, the more galaxies you can grab at a time, the better your chances of finding even one, and in identifying supernova candidates ,quantity still trumped quality. The photographic plates were big—eight inches by eight inches—and covered a patch of sky equal to one hundred full moons. For the follow-up observations, they would use the narrow-field CCD, a telescope with only a one-moon view, but that window was wide enough for performing

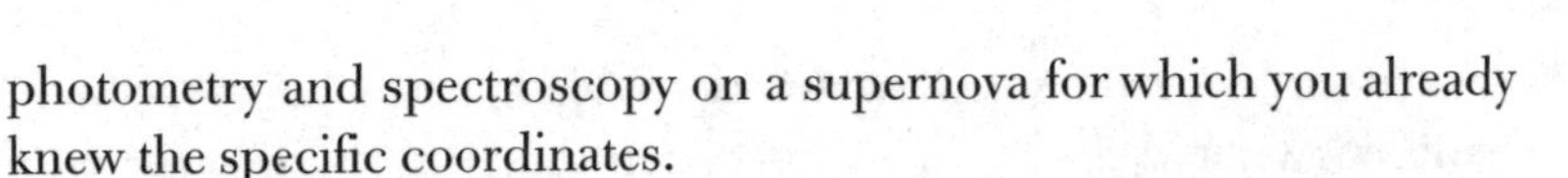

photometry and spectroscopy on a supernova for which you already knew the specific coordinates.

The workday for the Calán/Tololo collaboration would begin at sundown at the Curtis Schmidt Telescope, where the Cerro Tololo team would take the images and develop the photographic plates. At sunrise, they would put the plates on a truck, which would deliver the plates to a passenger bus, which would arrive, via the coastal highway, seven or eight hours later in Santiago. There the bus would be met by research assistants from Cerro Calán, who would bring the plates back to the observatory and then blink the previous night's images with reference images from a few weeks earlier. By the time the sun was setting and the dome was opening in Cerro Tololo, Hamuy, Phillips, and Suntzeff would have a list of supernova candidates they would be chasing that night with the CCD.

The survey, however, wouldn't just discover supernovae. It would also improve the field's quality of observation and analysis by following up on the supernova discoveries of other astronomers, both professional and amateur. The team looked at the two odd 1991 supernovae—the surprisingly bright 1991T and the surprisingly dim 1991bg; Phillips became the lead author on an *Astronomical Journal* paper analyzing 1991T. Those two supernovae only reinforced his and Suntzeff's earlier suspicion that supernovae weren't standard candles. You could see the disparity at a glance—the light curves were just that different. The light curve belonging to the surprisingly bright 1991T rose and fell more gradually than the typical Type Ia light curve. The light curve belonging to the surprisingly dim 1991bg rose and fell more abruptly than the typical Type Ia light curve.

The bright one declined more gradually. The dim one declined more abruptly.

Bright . . . gradually. Dim . . . abruptly.

The correlation jolted Phillips. Would it hold if he examined the light curves from a range of supernovae? If so, then maybe Type Ia didn't have to be identical in order to be useful for cosmology. Maybe how gradually or abruptly a light curve rose and fell could serve as a reliable indicator of its brightness relative to other Type Ia supernovae. And if you knew the relative luminosities among supernovae, then, through the inverse-square law, you would also be able to figure out relative distances. You would be able to use supernovae to do cosmology.

Phillips recalled that while working on 1986G—the first CCD supernova light curve—he had consulted papers by Yuri Pskovskii from 1977 and 1984 positing a relation between the rise and fall of light curves and their absolute luminosities. But Phillips knew that the uneven quality of photographic plates made Pskovskii's hypothesis untrustworthy. Now, Phillips figured, he could use nonphotographic studies to resolve the question.

Throughout 1992 he collected light curves, including some of his own, that he felt satisfied the most stringent observational criteria, and then he subjected them to months-long analysis. One morning late that year, he felt his preparations were done. The time had come to take the light curves, nine in all, and plot the data.

One of the advantages of living in a relatively small city like La Serena, Chile, Phillips thought, was that you could walk home and have lunch with your wife. He usually didn't discuss his work with her. She wasn't particularly interested in astronomy, and he didn't particularly feel like talking about it at home. But that afternoon he made an exception.

"I think," he told his wife, "I've discovered something important."

One of the first astronomers to write Phillips with congratulations was none other than Bob Kirshner—a blessing from afar, a benediction from above. The Danish and Berkeley teams had both asked whether one could discover supernovae at distances sufficient to do cosmology. Their answers, in 1988 and 1992, were: Yes. Now the Calán/Tololo team had taken what Kirshner considered the scientifically responsible first step and answered the question of whether Type Ia were standard candles: No. But they might be the next best thing: candles you could *standardize*. You could correlate the decline rate of the light curve with the supernova's absolute magnitude.

The next question, then, was: Could one detect distant Ia supernovae on a regular and reliable basis?

In March 1994, out of nowhere, the Berkeley team answered that question decisively: Yes. In a stunning announcement, they said that between December 1993 and February 1994 they had discovered six distant supernovae in as many nights.

When Schmidt sat down with Suntzeff in La Serena in late March and discussed the possibility of competing against Saul, the repercussions of the announcement were still reverberating in the supernova community, like one of the aftershocks perpetually roiling the Chilean countryside. Schmidt and Suntzeff themselves were reeling from a sickening realization. Since 1989, Calán/Tololo had collected fifty nearby supernovae, of which twenty-nine were Type Ia. The members of the collaboration would soon be publishing their value for the Hubble parameter, at which point their data would become public information, freely available. Eventually Berkeley was going to need nearby Type Ia to anchor the lower end of their Hubble diagram.

*Saul was going to be using their data to beat them at their own game.*

Extending the Type Ia search to higher redshifts, upon the completion of the nearby survey, had always been a possibility. Now it had become urgent. But they would have to move fast if they were going to collect enough distant supernovae of their own.

Brian Schmidt was the beneficiary of an oil-money education: a secondary school in Alaska that hired PhDs to teach teenagers. Schmidt hadn't been the best physics student in his class; he reckoned that two other students were superior. But he was the one who wound up putting physics to use, and he attributed the difference to "passion." If he said he could write the code for a high-redshift supernova search in two months, then he would write the code in two months. And he did, sort of. He wrote some new code and patched in some exisiting code from Phillips and Suntzeff's CCD observations of 1986G and on the basis of the resulting test "data," he and his team received three runs of two nights each on the 4-metre telescope at Cerro Tololo early the following year, in February and March 1995. By that time Schmidt was in the process of moving to Australia with his wife and their three-month-old — he would be living in Canberra and working at the Mount Stromlo and Siding Spring Observatories — and he didn't have travel money. But he figured he could manage the observations with the help of on-site collaborators and the Internet.

That's when he had his first epiphany: *Test data isn't the real thing.*

In February 1995 the team began taking the reference images — the initial images that they would subtract from later images. Sometimes the software wouldn't run. When it did run, Schmidt found himself trying to debug software and download images on a 100-byte-per-second link. Often he would have to guess what the problem was and do minuscule modifications of code, which his collaborators in Chile would implement, and then inevitably they'd all be talking on the phone, Cambridge and La Serena and Canberra, and Schmidt would wind up staying up all night while his wife, caring for the baby, hovered nearby. Finally Schmidt instructed his collaborators in Chile to send a set of data to him on tape so that he could examine it himself.

It never arrived. Somewhere between Santiago and Siding Spring, it vanished.

That's when he had his second epiphany: *From now on, I will always go to Chile.*

"From now on," though, presupposed that they would find a distant Type Ia. The first night of follow-up observations, 24/25 February, was clear, but the "seeing"—the term astronomers use for atmospheric conditions—was poor. The second night, 6/7 March, had excellent seeing and produced six candidates, but on closer inspection none turned out to be a supernova. The third night, 24/25 March, was, except for one brief period, overcast.

In the Observing Proposal that the team had submitted the previous September, Schmidt and Suntzeff had written, "Based on the statistics of discovery from the Calán/Tololo SN survey, we can expect to find about 3 SNe Ia per month." Surely, they told themselves now, they should find at least *one.* Yet they were heading into their final night, and they'd seen nothing. Were they doing something wrong? Did the software harbour a glitch? Or had they just been unlucky?

On the final night, 29/30 March, a 16-pixel-by-16-pixel smudge of light squeezed through the Internet pipeline between Chile and Australia. Schmidt squinted at it, then squinted at it some more, but he couldn't be positive. He picked up the phone and called Phillips and Suntzeff and told them to take a look and tell him what they thought. They took a look and agreed that it sort of seemed like a supernova, maybe. But they couldn't know for sure until they got the results of the crucial follow-up spectroscopy, from the 3.6-metre New Technology Telescope in La Silla, Chile.

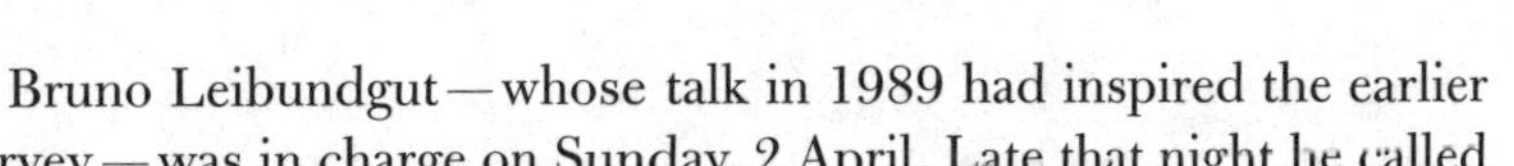


Bruno Leibundgut—whose talk in 1989 had inspired the earlier survey—was in charge on Sunday, 2 April. Late that night he called Mark Phillips with a bad-news/good-news report.

First the bad news. "It's very faint," he said.

But the good news was that at least the smudge was still there, indicating that it was indeed a supernova.

So: They'd been merely unlucky.

The team continued to sort through the data from the March runs, and by Wednesday of that week they had found that the candidate supernova from the final night had also appeared on the last field from the third night of observing, just before the clouds closed them down. "This is very encouraging," Phillips e-mailed the team.*

Thursday morning, Chile time, Schmidt sent out a progress report to all the members of the collaboration. He suggested they start thinking about submitting the observation to the International Astronomical Union circular, a standard procedure. He also suggested that they start thinking of what to call the collaboration—"a catchy name (or at least an accurate one if we cannot be catchy)." They were moving forward as a team, though he still wished, he added, that they had a redshift for the galaxy.

That evening, they did. Mario Hamuy had examined the spectra that Leibundgut had obtained four nights earlier and reported the result to Phillips, who relayed it to the rest of the collaboration: The host galaxy, and with it the supernova itself, showed a redshift of 0.48, placing it at a distance of 4.9 billion light-years—and setting a new supernova record.

They couldn't yet tell if it was a Type Ia, meaning that they didn't yet know if it would be useful in determining the rate of deceleration. Leibundgut would have to keep pounding at the data before they could say anything with confidence.

But as Leibundgut wrote to Schmidt that day, "We are rolling again. What a change a supernova can make."

They had beaten Saul at beating them at their own game.

---

* Quotations from e-mails throughout the book preserve the original spelling, capitalization, and punctuation.

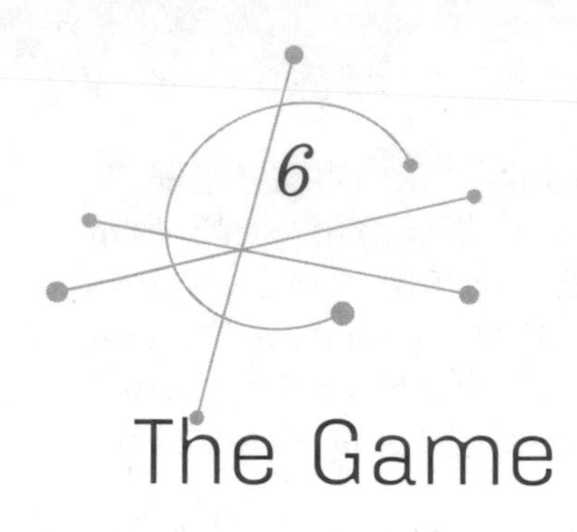


# The Game

POP! SN 1994F went off.

*Pop!* SN 1994G went off.

*Pop!* SN 1994H went off.

The Berkeley team had hung numbered tags around the necks of the champagne bottles, "1" through "6," one for each supernova they had discovered during their most recent run at the Isaac Newton Telescope, December 1993 through February 1994, plus a "0" for their 1992 supernova. The members of the Supernova Cosmology Project—as they were now calling themselves—had gathered at Gerson Goldhaber's house in the Berkeley hills. They laughed at themselves for being "lightweights"; they probably couldn't finish even two bottles. But the champagne inside the bottles wasn't the point, of course; the number of bottles was. It had taken the team four years to get their first supernova. Now, it had taken them three months to get their next six.

But they weren't celebrating only the supernovae. They were toasting their survival. Carl Pennypacker was no longer part of the team. Pushed? Jumped? Who knew? At least they still had a team. In the autumn of 1994 the Center for Particle Astrophysics and the LBL Physics Division had convened a Project Review Committee to determine whether the supernova search should continue. Even when the committee decided favorably, Bernard Sadoulet cut the CfPA's contribution to the supernova budget by half and gave it to his own project; Robert Cahn, the head of the LBL Physics Division, then informed Sadoulet that he was cutting LBL's contribution to Sadoulet's dark-matter experiment by half and giving it to the Supernova Cosmology Project. Not only was the SCP solvent for a change, but they had a guardian angel at LBL, a division director who understood why a physics lab might want to do a distant supernova search. And now they'd gone and found six more supernovae. They weren't just in the distant supernova game, they *were* the game.

*Pop!* SN 1994al. *Pop!* SN 1994am. *Pop!* SN 1994an.

The party didn't last long.

For years Bob Kirshner, as a member of the supernova project's External Advisory Board, had been saying that the LBL collaboration didn't know what it was doing—that team members weren't taking dust into account or paying sufficient attention to photometry or concerning themselves with whether Type Ia supernovae were standard candles. He didn't seem to understand that for LBL such considerations were beside the point—or weren't *yet* the point. The team was just hoping to prove it could do what it was trying to do: detect supernovae distant enough to be useful for doing cosmology. Their early efforts were what team members called, on various occasions, "demonstration runs" that were part of a "pilot search" in an "exploratory program." Then when they did find the 1992 supernova, Kirshner's objections as referee on that paper held up publication until 1995, when a more sympathetic referee, Allan Sandage, approved it. Breezing into Berkeley from Harvard, Kirshner seemed oblivious to the growing consternation, frustration, and anger not only at his objections but at *him.* A colleague of his on the External Advisory Board, speaking at a cosmology conference, characterized Kirshner's contribution to the discussion of the LBL approach: "No! This could not work! It couldn't possibly discover these high-redshift supernovae!"

And now here Kirshner was, saying, *Maybe.*

At least the LBL team had a six-year head start—surely that counted for something. Besides, their faith in Type Ia as standard candles *had* been rewarded. First Mark Phillips had demonstrated that the light curve for an inherently dimmer supernova falls off sooner—its descent is steeper—than the curve for an inherently brighter one. Then the Berkeley team had arrived at their own variation on his technique. They made Type Ia light curves uniform by treating them like images in a funfair mirror—stretching them "fatter" or compressing them "thinner" until they fit an idealized Type Ia template. (The team often took advantage of an LBL photocopier that could distort images in precisely that manner.) If Type Ia supernovae were less a type than a family, then each member of that family was less a standard candle than a calibrated candle.

And now, three years after proving to themselves that they could find a distant supernova, they had gone ahead and figured out how to find supernovae on a regular basis. After discovering the three in early 1994 on the Isaac Newton Telescope, they found three more with the Kitt Peak 4-metre telescope, in the mountains southwest of Tucson, Arizona. By June 1995 they had accumulated eleven distant Type Ia in total, and they were ready to make their first major statement to the community in the form of four papers to be presented at the NATO Advanced Study Institute Thermonuclear Supernovae Conference in Aiguablava, on the Mediterranean coast of Spain: They had figured out a way to discover Type Ia supernovae whenever they wanted.

They called it the "batch" method. Just after a new moon they would make as many as a hundred observations, each image containing hundreds of galaxies as well as, if possible, clusters of galaxies. In one several-day run they could gather tens of thousands of galaxies. Two and a half to three weeks later, just before the next new moon, they would return to those same tens of thousands of galaxies. In an update of the old galaxy-by-galaxy blinking technique, computer software would subtract the earlier reference image from each new image, searching the hundreds of galaxies for the new dot of light that might signal the emergence of a supernova. Then, again by using new software, the astronomers could determine *that same night* whether that dot actually was a supernova. They could then relay those coordinates to team members waiting at other telescopes, who would perform the necessary spectroscopy and photometry using telescopes on which they'd reserved time months earlier. (Having a successful track record worked wonders with time allocation committees.) You might not know in advance exactly where a supernova was going to go off, but you knew that one or more would. In effect, they had figured out how to "*schedule* supernova explosions"* —just like that.

*Pop, pop, pop.*

So Berkeley had a six-year head start. So what? Schmidt and Suntzeff's team had *astronomers* — professionals who didn't need to learn how to do photometry and spectroscopy, who needed only to do them well and then to make improvements where necessary.

A lot of Schmidt and Suntzeff's team had been at the NATO meeting, too. (The conference was organized by a former postdoc of Kirshner's.) The Harvard and Chile guys regarded the Berkeley team with some incredulity. "'I just heard about it, and I just thought about it, so — this is my subject!'" is how Kirshner characterized the SCP's attitude towards supernovae. SCP team members were talking about timing their observations to the new moon — as if astronomers hadn't been doing just that for thousands of years. The "batch" method? Radio astronomers were taking that approach in the 1960s. Supernovae on demand? José Maza was delivering that in the 1970s.

Naturally, with all these scientists pursuing the same goal, meeting in the same place, there had been talk of a collaboration. But some of the members of Schmidt and Suntzeff's team left Spain with the impression that, as Kirshner said, "working together meant working for them." Why would one of the world's most knowledgeable supernova specialists want to be subordinate to Saul Perlmutter, Type Ia neophyte and ten years his junior? For that matter, why would any of these purebred astronomers put themselves in a position where they might be reporting to purebred physicists? Perlmutter was talking about how "rare," "rapid," and "random" supernovae were. And they were! But Schmidt and Suntzeff's team preferred to put the emphasis on "dimness," "distance," and "dust" — how to tell whether a supernova is intrinsically dim, dim because it's distant, or dim because of dust. While the physicists were worrying about how to find distant supernovae, the astronomers were worrying about what to do with the distant supernovae once they found them.

Which they had (or, that is, *one* distant supernova). But they still

---

* Their italics.

couldn't be sure what type it was. The problem had been there from the start. In the same 6 April e-mail that Leibundgut sent to Schmidt celebrating the difference that one supernova can make, he also mentioned, almost as an aside, that "the 'supernova' spectrum still has a lot of galaxy in it"—that the light from the apparent supernova was difficult to separate from the light of the host galaxy. The spectrum could tell you the redshift of the galaxy, and therefore the redshift of the supernova residing in it. But in order to see the spectrum of the supernova itself, you were going to have to isolate its light.

First Mark Phillips tried. One week after notifying the team of Hamuy's calculation that the supernova was the most distant ever, he was ready to give up. "I've spent TOO MUCH time the last few days looking at this," he wrote the team. "The conclusion I've reached is that the SN spectrum is of such low S/N"—signal-to-noise, useful supernova light versus the optical equivalent of static from the galaxy—"that it is IMPOSSIBLE to tell what type it is."

Leibundgut tried next. And tried. "And no to the spectrum," he wrote in an e-mail at the end of May. "I have tried several ways of extraction but without any improvements." When he got to the Aiguablava conference, he told his collaborators that he, too, was ready to give up. "I don't know what to do anymore," he said. "I'm not sure I can confirm it's Type Ia."

"Crap!" Suntzeff said. "Saul's pulling in supernovae by the handful, and we only have one, and we can't even tell if it's a Ia!"

At one point Leibundgut was discussing the problem with Phillips in the lobby of the hotel. The sea was outside. They were inside. Phillips turned to Leibundgut and said, "Why don't you subtract the galaxy?"

"Subtracting the galaxy" is just about the first thing you do if you're trying to get the spectrum of a supernova. If you want to isolate the supernova light, you take a spectrum from the part of the galaxy containing the supernova, which is flooded with light from the galaxy, and then you take a spectrum from a different part of the galaxy, away from the supernova, and then you subtract the second reading from the first. Ideally, the spectrum of the supernova itself pops out.

This supernova, however, had been so overwhelmed by galaxy light that Leibundgut hadn't tried the obvious. Nobody had. From

Aiguablava he flew to Hawaii for another conference, and then home to Munich. He fiddled a little with the overall galaxy light, dividing its intensity by ten. Why ten? No reason. The spectrum from the galaxy would still be the same; he wasn't changing the quality of the data. He was just changing its intensity. He subtracted this spectrum from the supernova spectrum (which also contained the galaxy spectrum), and out popped a beautiful supernova spectrum.

"The spectrum of 95K looks great!" Phillips wrote to him on 1 August. "I'm now very convinced that this was a genuine type Ia."

They were back in the game. Now what they needed was to formalize their existence as a team.

From the start—during their first discussions in La Serena, in early 1994—Schmidt and Suntzeff knew what kind of team they would want. Theirs wouldn't resemble a particle physics collaboration. It wouldn't have the same rigid top-down hierarchy, the same plodding bureaucracy, the same assembly-line mentality. Instead, their collaboration would follow a traditional astronomy aesthetic. It would be as nimble, as independent, as Hubble on Mount Wilson or Sandage on Mount Palomar.

Already that approach had paid off. Like the professional astronomers they were, they had asked what they considered the key question first: Are Type Ia supernovae really standard candles? Only when they knew that Type Ia could be calibrated did they actually go looking for a distant one. And they almost hadn't found it. But in the end they did find their high-redshift supernova, and it was indeed a Type Ia. They'd salvaged their collaboration, and maybe their credibility, by making the discovery "on the smell of an oily rag in a quasi-chaotic fashion," as Schmidt liked to say. The process hadn't been pretty—it was, Suntzeff thought, more like "anarchy"—but it was astronomy.

And yet, astronomy itself was changing. The traditional go-it-alone aesthetic was disappearing. The diversity of the science and the complications of technology were forcing the field into greater and greater specialization. You couldn't just study the heavens anymore; you studied planets, or stars, or galaxies, or the Sun. But you didn't study just stars anymore, either; you studied only the stars that explode. And you didn't study just supernovae; you studied only one type. And you didn't study just Type Ia; you specialized in the mechanism leading to the thermonuclear explosion, or you specialized in what metals the explosion creates, or you specialized in how to use the light from the explosion to measure the deceleration of the expansion of the universe — how to perform the photometry or do the spectroscopy or write the code. A collaboration could easily become unwieldy.

Suntzeff and Schmidt recognized that their team would have to reflect the reality of increasing specialization. As the project evolved from the back of a sheet of computer paper in the spring of 1994 to actual astronomers chewing antacid tablets at observatories, they had to consider not only who would work hard but who possessed what areas of expertise and who had access to the right telescopes. The team's first foray into legitimacy, the September 1994 proposal for observing time at Cerro Tololo the following spring, cited twelve collaborators at five institutions on three continents. After the team confirmed that they'd found their first distant supernova during that run, in early April 1995, Schmidt sent around a reminder of who they were: fourteen astronomers at six institutions. The paper announcing the discovery in the *ESO Messenger* that autumn carried seventeen authors at seven institutions.

Yet even as the collaboration grew, Schmidt and Suntzeff wanted to preserve the dexterity afforded by old-fashioned astronomy — and to turn their familiarity with that tradition to their advantage. They were, after all, playing catch-up.

"We can only do it if we're fast," Suntzeff said. "The only way we're going to get this done is if we recruit as many young people as possible." Young astronomers. Postdocs. Graduate students.

They also wanted the collaboration to be fair. "I'm tired of seeing people get screwed by the system," Suntzeff said — the system where the postdoctoral fellow did the work and the senior astronomer who had a safe tenure would be first author, getting the credit and going to conferences while the postdoc wound up without a job.

By the time Schmidt and Suntzeff gathered their collaborators at Harvard in the late summer of 1995, they had formulated a strategy for delegating responsibilities in a way that would move the project forward quickly and fairly. Each semester, one of the sponsoring institutions — Harvard, or Cerro Tololo, or the European Southern Observatory, or the University of Washington — would be in charge of gathering the data from all the collaborators, reducing it, and preparing a paper for publication. And whoever did the most work on the paper would be the first author.

Unlike an anarchy, a democracy—even a revolutionary one—needs a leader. In one sense, Kirshner was the obvious choice. But he was also the embodiment of what Schmidt and Suntzeff wanted to avoid; in addition to *Go fast* and *Be fair*, they had framed a corollary: *No big guns.* Back in La Serena in the spring of 1994, when they had made the initial list of potential participants in the distant supernova search, they hadn't even included Kirshner. In his PhD thesis, Schmidt had used Type II supernovae to derive a Hubble constant of 60, then watched Kirshner crow about it at conferences Although they'd eventually recognized that sidelining one of the world's most prominent supernova authorities, and a mentor to many in the group, probably wouldn't be wise either scientifically or politically, Schmidt and Suntzeff remained wary of big-gun syndrome.

And with good reason. At the January 1994 AAS meeting in Washington, Mario Hamuy had given a talk about the Calán/Tololo survey; he was working on an elaboration of Mark Phillips's discovery of the relationship between light curves and absolute luminosity. Afterward, at Kirshner's invitation, Hamuy continued on to Cambridge to give a colloquium on the subject at the Center for Astrophysics. After the talk, a graduate student had invited Hamuy to his office. Adam Riess, not yet twenty-five, radiated the kind of confidence that comes from being the younger brother of two adoring sisters. When he got into the supernova game, he saw no reason why he shouldn't solve the biggest problem out there—how to standardize Type Ia's. Now he wanted to show Hamuy a technique he was developing. Like Phillips's method, Riess's light-curve shape (LCS) allowed you to determine the intrinsic brightness of a supernova; unlike Phillips's method, LCS also provided a statistical measure—a way to refine the margin of error. It quantified the quality of the result.

Hamuy examined it and told Riess he thought it was, as scientists say by way of praise, "robust."

Riess, however, said he had a problem. So far he hadn't been able to test the LCS method on real data. Could he see Hamuy's?

Hamuy hesitated. Your data was your data. Until you published it, it was yours and yours alone. But Riess was persistent, and Hamuy was a guest (at *Harvard,* of *Bob Kirshner*), and he relented. Hamuy agreed to show Riess his first thirteen light curves, though not before exacting a promise: Riess could use them only to test his technique, not as part of a paper about the technique.

A few weeks later, Hamuy got an e-mail from Riess. The technique was working. Riess was excited. Could he publish the results after all?

That, Hamuy reminded Riess, wasn't part of the deal. But again he relented, though not before exacting another promise: that Riess wouldn't publish his paper using Hamuy's data before Hamuy published his own paper on the thirteen supernovae. Riess would have to wait until Hamuy's paper had cleared the referee stage at the *Astronomical Journal.* When it did, in early September 1994, Hamuy let Harvard—meaning Riess and Kirshner, as well as William Press, who provided mathematical guidance—know that they were free to submit their own paper.

They did. But they submitted it to *Astrophysical Journal Letters,* a publication that, as its name suggests, traffics in shorter papers—and, therefore, briefer lead times.

Hamuy had to work hard to convince the *Astronomical Journal* to rush his own paper into print. In the end both papers appeared in January 1995. Both papers used the data to derive a value for the Hubble constant. And both arrived at a Hubble constant in the 60s—Hamuy's "62–67," and Riess's "67±7." Forevermore, Hamuy understood, the two papers would be cited side by side, as simultaneous publications.

"How could I be stupid enough to say okay?" Hamuy moaned to his colleagues in Chile. "'Mario! Mario! Mario!'" he wailed, mocking himself as much as Riess's entreaties.

Nick Suntzeff could see Kirshner's clumsy thumbprints all over the handling of the timing of the Riess et al. paper. Besides, he was already suffering from his own brush-with-astronomy-greatness fascination. Allan Sandage had encouraged Suntzeff to use CCD technology on Type Ia supernovae to find the Hubble parameter, and Suntzeff had helped his team do so, but the value the Calán/Tololo collaboration derived was on the "wrong" side of 60. Astronomers had estimated that the oldest stars in globular clusters were around sixteen to eighteen billion years old. A Hubble constant of 50 would correspond to a universe that was maybe twenty billion years old; a Hubble constant over 60 corresponded to a universe that was maybe ten billion years old — a universe younger than its oldest stars.

Suntzeff knew Sandage's reputation even as he was befriending him in the early 1980s. Everyone in astronomy knew Sandage's reputation. Even Sandage knew it. But he couldn't help himself; he took the Hubble parameter personally. He had inherited the programme from Hubble himself, he had pursued it for four decades, he had wrestled the value down from the ridiculous mid-three-digits to the realistic mid-two. In the 1970s Gérard de Vaucouleurs had taken it upon himself to challenge Sandage's methodology and assumptions, and he'd arrived at a Hubble constant of 100. Other astronomers had begun finding values that roughly split the difference between 50 and 100. Sandage wouldn't budge. The Hubble constant had to be less than 60, he insisted; the age of the universe demanded it. "The answer will come," Sandage once sneered, "when responsible people go to the telescope."

And now Suntzeff had joined the ranks of the irresponsible. He had received a note from Sandage, accusing him of having fallen prey to unsavory influences. Suntzeff tried to contact Sandage. Then Phillips tried. But Sandage was done with them.

Suntzeff, however, wasn't done with Sandage. Having helped derive a value for one of the two numbers in cosmology, he was now mounting an assault on the other. Suntzeff could tell himself that the "competition" with Sandage was in Sandage's head; it was just Uncle Allan being avuncular with a vengeance. He had always known that Sandage might one day turn on him, just as Sandage had turned on other acolytes and colleagues once he thought they'd turned on *him.* But this business with Hamuy and Kirshner was something else. It wasn't just personally disappointing. It was professionally dangerous.

Actually, the battle line, as Suntzeff saw it, wasn't Hamuy versus Kirshner. It was Calán/Tololo versus Kirshner. You could hardly blame Riess, an overall affable guy, a graduate student presumably wilting under the will of a powerful mentor. But Kirshner should have known better. Did know better. And didn't care. José Maza, the University of Chile astronomer who had served as Hamuy's mentor, resigned from the collaboration even before the initial observing run in February 1995. Hamuy himself, disgusted and disillusioned, decided that now would be a good time to go back to university for his PhD; he would be heading for the University of Arizona in the fall of 1995. Suntzeff's colleague at Cerro Tololo, Mark Phillips, adopted a "We have to get past this" attitude; Kirshner had served on the advisory board at Cerro Tololo, and it was Kirshner who told Phillips about 1986g, the supernova that had launched Phillips's and Suntzeff's careers in the supernova game. Yet even Phillips readily acknowledged feeling that what Kirshner had done was "improper."

And then it got worse, at least from the point of view of the Chilean part of the collaboration. Even before the Hamuy et al. and Riess et al. papers made their simultaneous appearances in January 1995, Riess and Kirshner had submitted another paper using Hamuy's data, this time to study the local motions of galaxies. The Calán/Tololo collaboration felt, as Suntzeff said, "as if blood was shooting out of our eyes." Shouldn't the guys at Harvard have known that it was a subject Hamuy was likely to pursue? Shouldn't they at least have contacted him and offered to collaborate?

And now, only a month after *that* paper appeared in the *Astrophysics Journal,* Suntzeff had to help decide whether Kirshner should lead the team he and Schmidt had created.

Suntzeff wouldn't be team leader; he had known that from the start. While he wanted to be sure that the Chilean contribution was recognized, he also understood the reality of his situation.

"I'm a staff astronomer in Chile," he told Schmidt. This kind of project would take a 100 percent commitment, and he already had a full-time job—and that job was in a place that left him "really isolated." But there was an even more important consideration, he argued: The post would need someone who could bridge both worlds—or both hemispheres, anyway. It needed Schmidt.

In terms of leading the team, Schmidt was equal to Kirshner in all ways except seniority. He'd helped found the group. He'd led the charge in Chile. Perhaps most important, he was no longer at Harvard; he'd moved to Australia earlier in 1995 (a fourth continent!).

And over the past several years, both as a postdoc and now on his own as an astronomer, he'd been in Chile often enough to know everyone there well and for everyone there to trust him.

Schmidt was reluctant. But he was also the guy who thought he could write code in two months.

"Yeah," he finally told Suntzeff, "I can do it."

Suntzeff campaigned quietly on Schmidt's behalf. Brian, he argued, had the personality to hold the group together, and he had the drive to get the job done. Eventually Suntzeff talked to just about everyone in the collaboration. Everyone except Kirshner.

Kirshner campaigned on his own behalf. His argument was that he knew the supernova game better than anyone. To a large extent, over the past quarter of a century he had *made* the supernova game what it was. He had a long history of writing proposals, securing support, keeping collaborators together. He reminded the team members that having all this young talent together in one place—the Harvard Center for Astrophysics—was "not an accident." He was the one who identified the promising graduate students; he was the one who hired the postdocs. "That's something," he told them, "that has to do with making a place where this subject is being done at the highest level."

The more he talked, the more he sounded like a big gun.

The team met in a seminar room in the basement of the Center for Astrophysics. Kirshner and Schmidt waited outside. After a short while, the door opened.

The Big Gun was out. The Young Turk was in.

Schmidt had learned his lesson: This time he went to Chile.

And not only did he go to Chile for the autumn 1995 observing season, he got there nearly a week early to test the new code he'd written. He immediately discovered that it didn't work. He was still at the mercy of the observatory's computers; if aspects of the computers had changed since he wrote his code back home in Australia, then he'd have to rewrite his code. The first day in Chile he worked ten hours trying to fix it. The second day he worked twelve or fourteen hours. Third day, sixteen or eighteen. Fourth day, twenty, and then twenty hours the next day, and twenty the day after that. When he started running a fever and having heart palpitations, Schmidt figured it was time to sleep.

Given the standard scheduling logistics at telescopes, his team had had to apply for time at Cerro Tololo the previous spring, even before they'd found their first distant supernova. If they hadn't discovered 1995K, who knows if they would have received the time? If they hadn't satisfied themselves that it was a Type Ia, who knows if they would be using the time, or at least using it to search for distant supernovae? But they were a true team now; they'd even put the idea of distant supernovae in their name: the High-z team (*z* being the symbol for redshift). It had all worked out, though when Schmidt had told the team via e-mail that they'd gotten time on the telescope, he'd added, "The bad news is that Perlmutter has more nights."

The two teams had applied for time during the same observing season, and the Time Allocation Committee at Cerro Tololo had taken the Solomonic approach of assigning the teams alternating nights. To make the situation even more awkward, one of Nick Suntzeff's duties at the observatory was to provide technical assistance to visiting observers. On nights that he wasn't participating in his own team's search for supernovae, he was, grimly, watching over Saul's. The objective astronomer in him found the Berkeley team's work "quite impressive." Personally, though, he could only shake his head and deliver his verdict to his collaborators: "They're well ahead of us."

The corks were still popping, one for each supernova, but now most of the champagne was running down the drain.

During that observing run at Cerro Tololo in the autumn of 1995, the SCP team discovered eleven more supernovae, in one run doubling the number they had gathered over the preceding three years. They had mastered the technique. Astronomers gathered data in Chile, forwarded it to their colleagues in Berkeley, who passed along the information to colleagues at the new 10-metre telescope at the W. M. Keck Observatory in Hawaii, where the team had already secured time because they knew, months in advance, that they would have supernovae to observe on that date. For the astronomers at the telescopes, the observations still contained drama: corrections to code, struggles with weather, decisions on what to target, bouts of diarrhoea, and, in

Chile, the occasional earthquake. But back in Berkeley, the overnight delivery of data was becoming routine. After all, in a universe full of billions of galaxies, stars were exploding all the time. Supernovae were out there by the thousands, by the millions, every night, waiting to be harvested. The Berkeley team had refined their collaboration, turning it into the kind of assembly-line operation that Alvarez and Muller had foreseen nearly two decades earlier. They were producing the intuitively paradoxical and once unthinkable: "supernovae on demand."

Somewhere in the universe, a civilization died. In Berkeley, they yawned.

In January 1996, at the AAS meeting in San Antonio, Saul Perlmutter sought out Robert Williams, the director of the Space Telescope Science Institute—the scheduling headquarters for the Hubble Space Telescope. Perlmutter wanted to talk about the "batch" method.

"I think with this technique," he said, "we now have the possibility, for the first time ever, of applying for HST time to follow up these very high-redshift supernovae." He explained that by now the SCP team had discovered twenty-two distant supernovae—mostly Type Ia—through the batch method. They had proven that they could predict the date they would find supernovae: whenever they got telescope time. And they could predict where: among whichever thousands of galaxies they chose to scour. They could guarantee the discovery of supernovae. The choice of when and where was now *theirs*, not the night sky's.

HST required just that level of certainty. It wasn't like earthbound telescopes. You couldn't just submit a proposal and six months later show up with a finding chart. The instrument required extremely complicated programming; you had to have your metaphorical finding chart in hand months in advance, with very little leeway for last-minute (actually, last-week) adjustments. Perlmutter's argument was that this kind of preparation was what the batch method allowed. The combination of confidence and specificity could meet the intricate dance of demands that came with booking time on the Space Telescope.

The logistical details would still be daunting, but they'd be worth it. The Hubble Space Telescope didn't see a lot; its field of view was minuscule compared with the old 5.1-metre or new 10-metre behemoths on terra firma. But what it saw, it saw with a clarity that no other telescope could approach. Through a CCD camera on an earthbound telescope, a very distant galaxy appeared as a smudge of pixels. Subtracting the light of the galaxy to isolate the light of the supernova was difficult work; witness the four months Leibundgut needed to figure out that the "very faint" 1995K was a Type Ia. The high resolution of HST, however, would make a supernova pop out of its host galaxy. Subtracting the light of the galaxy would be not only easier but far more precise.

That increase in photometric precision was crucial. At the time, it was perhaps the only justification for using HST on a supernova search. As everyone in astronomy knew, the purpose of HST was to perform science you could do only from space. The two distant supernova teams had proven that you could do their science on the ground—not as well as you could do it with HST, but you could do it nonetheless. What Perlmutter would need from Williams, then, was a slice of Director's Discretionary Time, a perk that routinely comes with the title of observatory director.

Williams said the idea sounded promising. He suggested that Perlmutter submit a proposal. A month later, Perlmutter did.

Williams, however, was no expert on the subject of distant supernova searches. Few people were, and with the exception of some Danes, nearly all of them were on one of the two competing teams. Three months later, at the annual May symposium at the institute, the SCP proposal came up during an open discussion. Bob Kirshner was in the audience. He had served on an HST Time Allocation Committee that had considered a similar proposal from SCP, and he had advised against it if only because the mission of HST was to do astronomy you could do only in space. He began to object, but Williams said they would continue the discussion in private. At the next break, Williams ushered Kirshner as well as Mark Phillips and Nick Suntzeff into his office.

"Is this a good idea?" he asked.

Kirshner immediately spoke up.

"No," he snapped. "It's the wrong idea." The point of a *space* telescope, he reminded Williams, was to do observations that you couldn't do on the ground. That was what the Observing Proposal paperwork said. That was what the previous STScI director had always insisted.

Williams listened. Then he said, "Yes, but I'm the director, and I can do what I want. This is really good science, and I think the Space Telescope ought to do anything that's really good."

Kirshner disagreed, and he and Williams went back and forth like this for a while. Occasionally Phillips and Suntzeff spoke up, echoing Kirshner's arguments. Still, the three High-z members knew what was at stake: If Perlmutter got HST time, that could well be the game. And clearly Williams wanted to give HST time to Perlmutter. He didn't want to hear the argument that nobody should be using HST to do follow-up photometry on distant supernovae. He wanted the best science to come out of HST. He wanted the best science to come out of *his* telescope.

Maybe they all realized it at once. Maybe they realized it one at a time. But at some point each of the three High-z members at the meeting understood what Williams was saying. If they asked for HST time, right then and there, they'd get it too.

They asked.

"My God, what an idiot!" Suntzeff thought as he left Williams's office. "Instead of pushing for the science that I want, I'm trying to argue, for moral reasons, about why we shouldn't be getting the data that we want! How stupid can you get?"

From the West Coast, the SCP watched agog. The dots weren't difficult to connect. Bob Williams had been director of Cerro Tololo from the mid-1980s to 1993. Mark Phillips and Nick Suntzeff had worked at Cerro Tololo under Williams. Bob Kirshner had served as an advisor to Cerro Tololo during this period. If you wanted to see evidence of an old boys' network, you didn't have to look very hard.

Bob Cahn, the director of the Physics Division at LBL, got on the phone with Kirshner and yelled for a bit. He got on the phone with Williams and yelled for a while.

Williams responded calmly, trying to explain his reasoning. HST was an important resource, and the search for high-redshift supernovae was a new field, and HST would surely get better results if both groups used it for their nearly identical experiments.

Cahn replied that he was familiar with important resources. He explained that high-energy physics, too, uses an important resource. He reminded Williams that this important resource was one that LBL had helped invent: the gigantic particle accelerator. But when a group applied for time on a gigantic particle accelerator, the proposal was confidential. Wasn't that how astronomy worked?

Williams conceded that that was indeed how astronomy worked, usually. He suggested a compromise. Both teams would receive Director's Discretionary Time, and the SCP would get to go first.

Cahn and Perlmutter had no choice but to accept. Afterwards, though, whenever SCP team members talked among themselves about their rivals, they did so with a new appreciation of just how formidable they were. In the culture of high-energy physics, scientists have to work in large collaborations, and those collaborations have to endure for a long time. You can't afford to alienate your competitors, if only because they'll soon be your collaborators. Astronomers, however, still roamed some Wild West of the mind, where resources were scarce, competition was fierce, and survival depended on small alliances of convenience, often enduring just long enough to publish a paper. Astronomers could afford to play for keeps.

Not that high-energy particle physicists weren't competitive. But in the end they had to get on if they wanted to get work done. They could be tough. But next to astronomers, they were, said one SCP partisan, "pussycats."

By the autumn of 1997, the two teams had enough data to try to find at least a preliminary answer to how much the rate of expansion of the universe was slowing down, and therefore whether the universe was heading towards a Big Crunch or a Big Chill.

As part of the High-z team's fast-and-fair philosophy, Schmidt had divvied up the responsibilities not only institution by institution but junior astronomer by junior astronomer. In this game of tag, the Australian National University Mount Stromlo and Siding Spring Observatories' Schmidt was "it" first. He would write up the paper broadly introducing the collaboration's methods and goals. Then the team tagged Harvard and Peter Garnavich; he would take the three Type Ia supernovae that the team had measured photometrically with HST in the spring of 1997, add 1995K, and try to figure out a value for the

Hubble constant. Just as important, the paper would help justify requests for more HST time.

On the SCP side, no one person was working exclusively on the problem. In keeping with the particle physics culture, the team was moving forward collectively. In fact, they'd already moved forward; a year earlier they announced the results from the first seven Type Ia, which suggested that the universe was flat—neither expanding forever nor eventually contracting. But the margins of error and the size of the sample were such that the result was preliminary at best.

Or wrong, as the team was beginning to suspect, on the basis of the one supernova on which they had managed to conduct reliable HST photometry. That one "guy," as astronomers like to call a piece of evidence, was indicating a possible shift in another direction, towards an open universe.

One approach astronomers could use in trying to make this determination was a histogram. On the morning of 24 September, Gerson Goldhaber sat at his desk at LBL to prepare for the weekly team meeting. Unlike a graph, which plots each individual point of data, a histogram gathers several pieces of data at a time and "bins" them in categories. That morning, Goldhaber took each of the thirty-eight SCP supernovae so far and, based on its brightness and redshift, put it in a column corresponding to the amount of matter that this one supernova suggested the universe needed in order to slow the expansion to a halt: 0 to 20 percent of the necessary mass density, 20 to 40 percent, and so on, up to 100 percent. When he was done, the two tallest columns by far, one of them bulging with ten supernovae and the other with nine—half the total sample—told him that not only did the universe not have enough matter to slow the expansion to a halt, it had 0 to *negative* 40 percent.

"Lo and behold," he said to himself.

For the High-z team, Adam Riess was working on a statistical approach to the problem. His task was to take all the supernova data collected so far—all the pixels of spectroscopy and photometry, all the galaxy subtractions, all the light curves, all the margins of error—and develop software that would compare it with millions of different models of the universe. Some of those models would be absurd: relationships between magnitudes and redshifts that, on a sheet of graph paper, would fall far from the straight, 45-degree line, off to the remote corners where the punch holes were punched. Other models, however, would match slight deviations from that seemingly straight line. Within this subset, some models would match even slighter deviations, even subtler departures from the "norm." One of those universes would match his data.

And one did. It was a universe that not only didn't have enough matter to slow the expansion but had a mass density of *negative* 36 percent. It was a universe without matter. It was a universe that didn't exist.

"Lo and behold," Riess told himself.

Both teams had been operating under the assumption that the universe was full of matter and only matter. They knew some of it was dark, of course, but what was missing was still fundamentally matter. They had therefore assumed that only matter would be influencing the expansion of the universe.

Abandon those assumptions, however, and these seemingly nonsensical results might make sense after all. If the two teams considered a universe in which something else was affecting the expansion —a universe that consisted of something other than matter—then the universe would have matter in it again. They looked at the error bars and figured that the matter, dark or otherwise, was maybe 20 or 30 or 40 percent. Which left 60 or 70 or 80 percent . . . something else.

As for the fate of the universe: They had their answer. Maybe even *the* answer—one they could quantify: It would expand forever.

What they didn't have—between the dark matter they couldn't see and this new force they couldn't imagine—was any idea what the universe was.

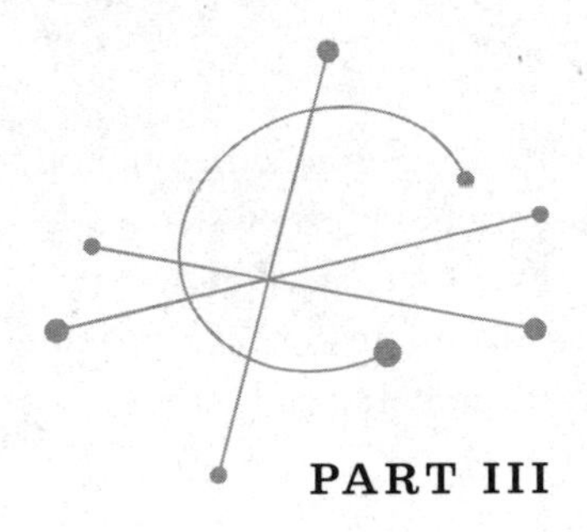

PART III

# The Face of the Deep

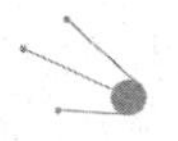

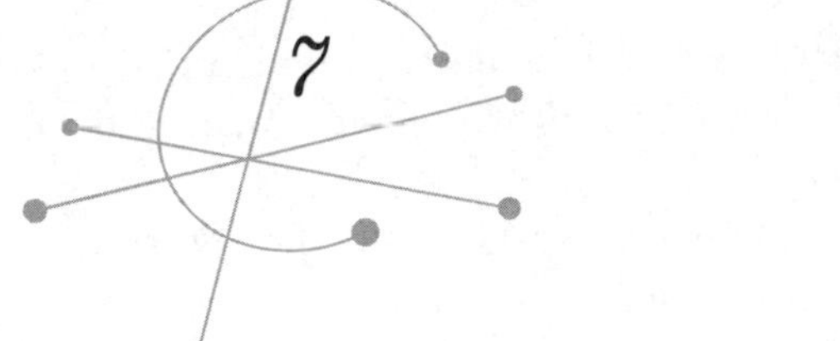


# The Flat Universe Society

ON MONDAY EVENINGS throughout the mid-1980s, the DuPage County Center for Scientific Culture held what would have been the only course in its catalogue, if it had had a catalogue. The classroom was the basement of a split-level suburban home. The student body was sparse: a handful of researchers, postdocs, and graduate students from the University of Chicago or the nearby Fermi National Accelerator Laboratory, as well as, often, a distinguished visitor. The students served as the instructors, too. Tuition was five dollars a week, which bought you pizza (or sometimes "backup" barbecued hamburgers, resurrected from the bowels of the freezer), beer, and a turn at the blackboard.

The topics varied from week to week, and from moment to moment. The first topic of the evening might be a recently published paper that had gotten it all wrong, whatever "it" was, or a wildly speculative hypothesis that someone wanted to test. From there the evening would follow its own path. The chalk would pass from hand to hand, feverishly, amidst shouts of criticism or approval and screams of sudden insight or instant regret, and by the end of the class the participants were vowing to write a response eviscerating the recent paper that had gotten it all wrong, whatever "it" was, or a new paper championing an original theory that, whether one of the participants had arrived at the meeting espousing it or it had arisen over the course of the evening, had already gone through its own peer evisceration. (Eventually the centre instituted a two- or three-day cooling-off period before participants could write up their papers.) But whatever path the evening had eventually followed, and however circuitously and riotously, the topic was always basically the same—what to do next with the Big Bang universe.

That universe was now nearly twenty years old. While observers were trying to measure the two numbers in cosmology—the universe's current rate of expansion, and how much the expansion was slowing down—theorists were trying to figure out how the expansion itself worked. Like Jim Peebles in his instant classic *Physical Cosmology*, they wanted to make explicit the connection between the physics of the early universe and the universe we see today.

That connection had been implicit from the start, in Lemaître's invocation of a primeval atom. And over the decades other theorists had tried to work out the calculations that would reveal how the universe had gotten from there to here—from hypotheses about a primeval fireball to observations of today's galaxies. The discovery of the cosmic microwave background, however, made a dialogue between particle physicists and astronomers *necessary*.

When the Princeton physicists had visited Holmdel in early 1965 to inquire about the detection of a 3 K signal, the Bell Labs astronomers explained what wavelength they had designed their antenna to detect, how they had taken into account the rattling of electrons—topics the Princeton physicists knew well. Their colleague Jim Peebles had already performed the calculation for the relic temperature of the primeval fireball, and Bob Dicke himself had invented some of the equipment in the Bell Labs experiment. Then the Bell Labs astronomers listened as the Princeton physicists talked about the Big Bang and the Steady State theories, how Dicke was hoping for evidence of an oscillating universe—topics the Bell Labs astronomers understood. Arno Penzias and Bob Wilson, like most astronomers, didn't take sides in the debate between the two theories, though Wilson had studied with Fred Hoyle and felt a slight allegiance to a Steady State universe. But that summit on Crawford Hill nonetheless marked the moment that particle physicists and astronomers began to talk to each other in earnest, with a sense that the conversation might actually lead somewhere: from here to there—from the current constitution of the universe to finer and finer fractions of a second in its history.

Hence the nickname for the nonexistent DuPage County Center

for Scientific Culture: Primordial Pizza. The real institution was the NASA/Fermilab Astrophysics Center, five minutes away, and Edward "Rocky" Kolb and Michael Turner had been hired to run the Fermilab centre in part because they were willing to entertain the unorthodox. Because Primordial Pizza met on Mondays, lectures sometimes fell on national holidays. No matter. Invitations to lead a seminar went out to distinguished visitors, who had to wonder why they were being summoned to Fermilab on Memorial Day, and who, after leading a seminar in the scholarly and respectful confines of a Fermilab conference room or auditorium, soon found themselves sitting in the "less than elegant surroundings" of a bachelor pad while being bombarded with questions by students dozens of years their junior.

Kolb had a wife and three kids, so to Turner went the honour of playing host. In the tradition of memorable comedic pairings — Laurel and Hardy, Morecambe and Wise, Cheech and Chong (Turner's reference of choice) — they complemented each other stylistically while possessing the same comic, and cosmic, sensibility. Kolb played strait-laced family guy, the tall lug with a push-broom moustache; Turner handled eye-rolling bomb thrower, long of hair and short of patience.

Kolb and Turner had both passed through Caltech — Turner as an undergraduate, Kolb as a postdoc. Even at informal meetings there, they found, you had to prepare meticulously, anticipating every possible objection. You were afraid to be wrong. Up the California coast, Luis Alvarez famously hosted biscuits-and-beer gatherings at "The Castle," his estate in the Berkeley hills. Each week a graduate student or postdoc had to present as-yet-unpublished news from the physics community. "I don't believe that," Alvarez would snap, moments into the talk, or "That doesn't make sense," or "The error bar doesn't look right." You had to explain and defend the research as if it were your own, even if you'd actually found it by ringing friends at Columbia or Harvard and begging them to throw you a scrap. And Stanford, where Turner had been a graduate student, "butchers its young." At those institutions, preparation was everything.

At the DuPage County Center for Scientific Culture, however, preparation was nothing. To prepare a presentation for Primordial Pizza was to violate its most solemn and sacred tenet: *Don't be afraid to be wrong.* And that directive went for graduate students and visiting Nobel laureates alike. The riff was more important than the result. You got up and improvised. You jammed. You played cosmology as if it were jazz.

Turner had inherited that sensibility, he had come to realize, from the bongo-beating quantum theorist Richard Feynman at Caltech — even though Feynman was, as Turner had also come to realize, "the worst advisor." Sometimes Feynman would advise graduate students to pursue subjects that, while of interest to him, would turn out to be beyond their understanding, and they wouldn't be able to complete their theses; sometimes he would advise PhD candidates to pursue subjects that — while, again, of interest to him — would turn out to be so obscure that they wouldn't be able to find postdoctoral fellowships. It was Feynman who had advised Turner to pursue his graduate studies at Stanford. Not until Turner got to Palo Alto did he realize that Feynman had suggested Stanford because it was the home of a linear accelerator that was performing particle physics experiments that were of interest to Feynman. "Feynman," Turner thought, "is interested in what Feynman is interested in."

What was Turner interested in? He didn't know. He was sharing a flat with some medical students, and he had to ask himself, What What was solving equations next to saving lives? He soon left graduate school and became a car mechanic. Earned $500 per study in drug experiments (marijuana, Valium) at the local military hospital. Worked weekends cleaning up after the one thousand animals in Stanford's research labs. If nothing else, these experiences instilled in him a deeper appreciation for the life he'd left behind — the life of the mind. And so, in time, Turner attended — he sat in the back of and took notes on, anyway — a course on general relativity.

General relativity wasn't quite right for him either. But at least the lecture found him back into the classroom, and back into physics. After Turner finished his dissertation, in 1978, the University of Chicago astrophysicist David Schramm called him with a postdoc offer. A few years earlier Schramm had found inspiration in *Physical Cosmology*, and since then he had been trying to yoke together the two topics that, individually, hadn't quite captured Turner's attention: particle physics and cosmology. Now Schramm said to Turner, in the same offhand manner that Bob Dicke had used with Jim Peebles when he suggested that Jim figure out the temperature for the cosmic microwave background, "Why don't you think about it?"

Turner talked it over with his PhD thesis advisor at Stanford, the physicist Robert Wagoner. "That early-universe cosmology stuff? Don't do that," Wagoner told him. Wagoner himself had participated in the Big Bang revolution. As a postdoc at Stanford in the two years immediately following the discovery of the cosmic microwave background, he had worked on the same kind of primordial particle physics that Schramm had adopted in the following decade. But he had a point. By the late 1970s, the Big Bang bandwagon had stalled. It lacked the one thing that could save it from swinging backwards to the voodoo side of the metaphysics–physics continuum, the one thing any theory needs to be scientific is a prediction to verify or falsify.

Whatever Feynman's liabilities as an advisor, he had taught Turner this lesson: Don't try to solve a problem until you think you have the answer. That approach was the opposite of how particle physics usually worked. In particle physics, the maths came first. The maths told you that a particle should exist, and that you could create that hypothetical particle from existing particles. Then you (and a thousand colleagues) commandeered an accelerator and smashed those existing particles together at velocities approaching the speed of light and waited for the hypothetical particle to pop into existence.

Nothing wrong with that approach. It worked.

But Feynman had taught Turner that sometimes you didn't need to do the maths first. Instead, you needed to trust your intuition. To leap to a conclusion first. To imagine what the universe might be, and then go back and do the maths until, with luck, it matched.

To imagine what your life might be, and then go back and do the work until, with luck, it matched.

"Don't do that early-universe stuff," said his thesis advisor at Stanford. "Come to Chicago and do great things!" boomed Schramm.

Schramm made you think you could do great things. He was, in a way, the embodiment of cosmology: big and bold and fearless. His colleagues calledhim Schrambo or Big Dave. At 6 feet 4 inches and 16 stone 6 pounds, he had the build of a former wrestler (which he was) and the bearing of an amateur pilot (ditto): king and conqueror of all he surveyed. When he decided to pursue the physics of the early universe during a period when the Big Bang was unfashionable, he not only made no apologies but claimed the field as his own. Big Bang Aviation, he named the corporation that controlled his private plane, of which he was sole proprietor. BIG BANG read the personalised licence plates on his red Porsche.

Turner might not have responded to cosmology or particle physics in isolation, but the combination proved irresistible—a balance of the loud, wildly speculative, and the quiet, "neat and simple." In particle astrophysics, Turner could reconcile the two dominant tendencies of his life. The bohemian who dropped out, the intellectual who crept back. The incautious and the careful.

So Michael Turner would go to Chicago. And he might even get to do great things—just as long as cosmology came up with a prediction.

In October 1981, Golden Tickets appeared in the letterboxes of cosmologists around the world, only the wonderland they would be entering at the appointed day and hour wasn't Willy Wonka's Chocolate Factory but Stephen Hawking's Nuffield workshop. The Nuffield Foundation, a charitable trust, had agreed to endow an annual workshop for three years. In the second year, Hawking and Gary W. Gibbons, also at Cambridge, decided to consolidate the remaining funds and go all out: an assault on the farthest frontier of cosmology, the "very early Universe," which the invitation defined as "< 1 sec."

Among the three dozen or so theorists who received the letter was Turner. He figured that Hawking and Gibbons had known to invite him because one of his colleagues at the University of Chicago was one of Hawking's frequent collaborators. Not that there were all that many theorists working this particular corner of the cosmological landscape. And of course a good workshop should have a fair number of young minds to ward against the calcification of old ideas and received wisdom. But Turner would have to earn his ticket, too. He would be one of a handful of attendees who would be not just giving a talk but writing a paper.

On his arrival at Cambridge on the first day of summer 1982, Turner presented the preliminary draft of his paper to Hawking.

Hawking nodded his thanks, then motioned to an assistant, who presented Turner with Hawking's paper. A couple of other papers were circulating as well. The time had come to confront a question that had been haunting cosmology since the day that Einstein extended general relativity to the universe: Why was the universe simple?

As the letter from Hawking and Gibbons had said, Big Bang cosmology "assumes certain initial conditions." Those assumptions, however, were notoriously ad hoc, from the Latin for *for this.* As in: *For this* purpose—the creation of a cosmological model from the general theory of relativity—Einstein assumed homogeneity, that the universe looked the same on the largest scale. *For this* purpose—the creation of a cosmological model that wasn't static—other theorists had added the assumption of isotropy, that the universe looked the same in every direction.

And the universe did seem to be homogeneous and isotropic. The discovery of the cosmic microwave background seventeen years earlier had satisfied most cosmologists that they now had the answer to the question of whether the universe was simple: Yes. On the largest scale it would look the same no matter where you were in it. And they had answered the question of *how* simple the universe was: Very. The cosmic microwave background was extremely smooth, just as theory had predicted.

But assuming that something is the way it is—even if those assumptions turn out to be correct, as the Big Bang theory's apparently were—is no substitute for understanding how it got that way. Why would a universe be, of all things a universe could be, simple—and not just simple, but *so* simple? On reflection, maybe the answer to the question of how simple the universe was shouldn't have been the satisfying "Very" but a suspect "Too."

Now, however, cosmology had a possible answer to the question of how the universe became so simple. Late in the evening of 6 December 1979, a no-longer-young academic with a boyish mop of hair, a boyish smile, and a grown-up worry about meeting the monthly rent sat down at the desk in his study, as he often did at that hour of the day. On this occasion, however, Alan Guth received a midnight visit from the Muse of Maths. The next morning he bicycled to his office at the Stanford Linear Accelerator Center (in the process establishing a new personal best of nine minutes and thirty-two seconds) and immediately sat down with his notebook to summarize his long night's work.

"SPECTACULAR REALIZATION," he wrote near the top of a fresh page, and then he did something he'd never done before with a notebook entry. He drew two boxes around it.

By the time of the Nuffield workshop, two and a half years later, the story was already the stuff of scientific legend. Guth had experienced a genuine "Eureka!" moment. His was the kind of insight that causes colleagues to slap their foreheads and groan, "Of course!" The day after Guth gave his first seminar on his spectacular realization, in January 1980, he received calls from seven institutions either inviting him to give the same seminar or asking if he would consider a faculty position. By then Guth had given his idea the name that stuck: inflation, a pun that accommodated both the defining physical property of his discovery and the dominant economic worry of the era.*

According to his calculations, the universe had gone through a monumental expansion in its first moment of existence. Aged just a trillionth of a trillionth of a trillionth of one second—or 1/1,000,000,000,000,000,000,000,000,000,000,000,000th of a second—the universe had expanded ten septillion-fold—or to 10,000,000,000,000,000,000,000,000 times its previous size.

The suggestion followed an idea that another physicist, Edward P. Tryon, had put forward several years earlier, in a 1973 article in *Nature.* Like Gamow with "Rotating Universe?"—the *Nature* article that had partly inspired Vera Rubin's master's thesis—Tryon couched the counterintuitive in the form of a question: "Is the Universe a Quantum Fluctuation?" According to the laws of quantum mechanics, virtual particles can arise out of the emptiness of space—and actually do, as experiments since the middle of the century had shown again and again. Tryon wondered if the universe might be the result of one such quantum pop.

---

*Guth, the son of a New Jersey grocer-turned-dry-cleaner who always seemed on the verge of going out of business, was partial to economic considerations. His insight had possibly not only saved cosmology but salvaged a career that was already on its fourth postdoctoral fellowship.

The argument became less sensational if you kept in mind that in quantum theory everything was a matter of probabilities. Therefore, anything was possible. Perhaps specific events were vanishingly unlikely—the creation of a universe from the nothingness of the vacuum, for instance. But they weren't impossible. And over the course of eternity, why shouldn't one or another of those vanishingly unlikely events come to pass? The universe, Tryon wrote, "is simply one of those things which happen from time to time." Or, as Guth liked to say, "the universe is the ultimate free lunch."

The problem with Tryon's idea was that it couldn't account for the size of our universe. Inflation, however, could. Guth realized that the infant universe could have gone through a process that physicists call a "phase transition" and everybody else calls "the thing that happens when water turns into ice or vice versa." When the temperature of water changes, the transformation doesn't happen all at once. It's not as if the word goes out and suddenly every molecule of $H_2O$ in the lake has melted into liquid or hardened into ice. Instead, the transformation happens piecemeal. Even within small sections of the pond the ice isn't freezing or melting uniformly. Cracks and fissures appear faintly, then harden, leaving a veined appearance. Guth found that if you apply that transformation mathematically to the conditions of the early universe, the phase transition would have produced a temporary vacuum. That vacuum, in turn, would have produced a negative pressure—a strong gravitational repulsion—that would have expanded space exponentially. The universe would have doubled in size, then doubled in size again, then doubled in size yet again. It would have done this at least a hundred times, and it would have done so over the course of $10^{-35}$ seconds (or $1/10^{35}$). After that, the vacuum would have decayed, the exponential expansion would have stopped, and the standard expansion of the universe—the one in the Big Bang theory that we can see for ourselves in the redshifting of the light from distant galaxies—would have begun.

At once Guth recalled a lecture by Bob Dicke that he had attended the previous year, one of a series that Dicke and Peebles had been delivering on a topic they called the "flatness problem." They would explain to their audiences that the fate of the universe depended on how much matter was in the universe: enough to reverse the expansion, not enough, or just right. The designation that scientists had given to the measure determining the fate of the universe was, aptly, the final letter in the Greek alphabet, omega. If the universe contained half the mass necessary to halt the expansion, then you would say omega equaled 0.5, or if it contained three-quarters of the necessary mass, you would say omega equaled 0.75. If the universe contained more than enough mass to halt the expansion, then omega equaled more than 1—1.5 times, maybe, or 2 times, or 100 times. And if the universe contained just the right amount—precisely the critical density to stop the universe from expanding but keep it from collapsing back upon itself—then omega equaled 1.

Astronomers would even be able to measure omega, if they had a standard candle that they could trace far enough across the universe. But you might not need observations to know omega, Dicke argued. Theory alone might be enough.

According to Dicke, any significant deviation from 1 in the earliest universe would have led, effectively and almost immediately, to the end of the universe: either an exponential expansion towards infinity or a collapse. Calculating backwards, the closer and closer you got to the Big Bang, the closer and closer omega must have been to 1. At three minutes after the Big Bang, omega would have been within a hundred-trillionth of 1. At one second after the Big Bang, omega would have been within a quadrillionth of 1—that is, between 0.999999999999999 and 1.000000000000001. The earlier in the universe you calculated, the more decimal places you added. At some point in the calculations you simply conceded: Omega as good as equaled 1.

And if omega equaled 1 then, it had to equal 1 now, because the value of omega depended on the measure of matter, and whatever matter the universe had then, it would have now and forever.

But for Big Bang theorists like Dicke and Peebles, a flat universe posed a problem similar to the one Newton and Einstein faced: Why would a universe that was full of matter not be collapsing through the effects of gravity? Newton had to invoke a universe of evenly spaced stars—plus God. Einstein had to invoke a universe of randomly spaced stars—plus lambda. Evidence for an expanding universe had allowed Einstein to abandon lambda and prompted future generations to try to figure how to measure the rate at which the expansion was slowing. But now Dicke and Peebles were arguing that in a Big Bang universe, omega had to equal 1. The expansion had to slow to a virtual stop and stay there forever. All the matter in the universe had to reach a state of gravitational equilibrium—an eventuality with the same likelihood as a pencil standing on its point forever. Not impossible, according to the laws of classical physics, but not likely either.

But that December night in 1979, Guth realized that if inflation did occur and the entire universe was actually ten septillion times the volume of what we see, then what was flat wasn't the universe but only our part of it—the part we'd always assumed to be the universe in its entirety. Our part of the universe would appear flat to us, just as a football field appears flat even though the Earth is round. The universe *as a whole* could have any value of omega; the universe that we *see,* however, has a value close enough to 1 that, for all practical purposes, it may as well be 1.

So much for the flatness problem.

A few weeks after inventing inflation, Guth was listening to some colleagues at lunch and learned about another apparent contradiction in a homogeneous and isotropic universe, the "horizon problem." Look into the universe in one direction, then look into the universe in the opposite direction. This is essentially what antennas measuring the cosmic microwave background do. The light from one direction will just be reaching you, and the light from the other direction will just be reaching you, but the light from the first source will not yet have had time to reach the second source, and vice versa. Yet the cosmic microwave background reveals a similarity in temperature to within one part in 100,000. How did one part of the universe "know" the temperature of the other and match it if the two had never "communicated" with each other?

"Hmm," Guth thought, "inflation could solve that, too." If inflation did occur, then two distant parts of the universe would have been in contact with each other when the universe was less than $10^{-35}$ seconds old. Guth thought a little more. Then he told himself, "This really might be a good idea after all."

Guth's paper "The Inflationary Universe: A Possible Solution to the Horizon and Flatness Problems" appeared in early 1981. While the Nuffield workshop was officially called "The Very Early Universe," it quickly became a referendum on inflation. Seventeen of the thirty-six sessions addressed the topic directly, and many of the others mentioned it.

The question wasn't so much whether inflation made sense. Inflation explained two ad hoc assumptions—homogeneity and isotropy. It solved two problems—flatness and the horizon. It was too good not to be true—or at least that's how many of the theorists at Nuffield felt. The question instead was whether they could fix its flaws.

Guth's original idea was plagued by a problem that he himself hadn't identified. Once his version of inflation started, it couldn't stop. Other theorists—Andrei Linde, at the Lebedev Physical Institute in Moscow, and, independently, Paul Steinhardt and Andreas Albrecht, at the University of Pennsylvania—identified the problem and found the solution. They reconceived the inflationary period to be, as Guth came to think of it, less like the bubbling of boiling water than the congealing of a single bubble in gelatin. The problem with the one-bubble inflationary model, however, was that it still had to account for the visible universe—homogeneous and isotropic, but not *too* homogeneous and isotropic, or else we wouldn't be here.

They were all borrowing from Hawking. In 1973 Hawking had redefined the study of the early universe with his work on black holes; he found that, owing to a combination of quantum and gravitational effects, they weren't one-way tickets to a singularity. At the edge of the event horizon—the black hole's ring of no return—quantum effects dictated that particles and antiparticles would be popping into existence, while gravitational effects dictated that one partner would disappear into the black hole but not the other. Rather than annihilating each other "immediately," one would slip over the edge, into the black hole, but the other would escape into space and the universe as we know it. Black holes, Hawking contended, aren't black after all. They leak radiation—Hawking radiation, as it came to be called.

In effect, Hawking had begun to bridge the two seemingly irreconcilable theories of the twentieth century, quantum mechanics and general relativity, a necessary step if science was ever going to describe the earliest, foamiest time after the singularity, or perhaps even the singularity itself. Two years later, Hawking and Gibbons extended the concept of quantum gravity to the universe as a whole and found that it would fill with thermal fluctuations. In early 1982, in the months leading up to Nuffield, Turner and Paul Steinhardt had begun working on the idea that those fluctuations could have been present during the inflationary period.

For Guth and Turner and some of the other attendees, Nuffield was the latest stop on what they'd come to regard as a "traveling circus" of cosmologists. In early 1982 they had attended conferences in London, the French Alps, and Switzerland. In April, Steinhardt and Hawking happened to be visiting the University of Chicago at the same time; head-scratching with Turner inevitably commenced. In May, Steinhardt visited Harvard; Guth biked over from MIT. In June, just two weeks before Nuffield, Hawking gave a lecture at Princeton; Steinhardt drove over from Philadelphia, then called Guth and Turner with the latest update on new inflation.

The conference at Nuffield was like that, only more condensed, more intense. Gibbons and Hawking had limited the schedule to two seminars most days, leaving the rest of the time for "informal discussion." And discuss they did. The participants talked during day trips to London. They talked over croquet and tea on the lawn at Hawking's house. They talked long into the night, knocking on one another's doors. And as they talked, the conception of inflation shifted, and shifted, and shifted again. For Turner, Nuffield was shaping up to be one of those rare cosmological events: "a workshop where work actually got done."

During his own talk, Turner tried to capture the breakneck exchange of ideas by adopting the tone of a TV newsreader and recapping the workshop so far in a series of "this-just-in" bulletins. He got laughs, but he also made a point: Even being able to analyse the problem was progress of a type. Had they made new inflation work? No. But they had agreed on a way that they might make it work. Now they knew they had the right equations, even if they hadn't yet figured out how to solve them. They'd managed to wrest the universe back from being "too" simple — ad hoc simple — to being merely "very" simple. They'd even managed to convince themselves that cosmology came with a prediction: The universe was flat.

"Child's play," Turner thought. Now came the hard part — and the fun.

What was inflation — what was Nuffield — if not an exercise in a Feynman kind of faith? Cosmologists in the early 1980s had leaped to a conclusion, embracing inflation simply because it explained and solved so much, and then they had gone back and laboured to make the maths work. And they'd succeeded. In the weeks following Nuffield, Turner and the other attendees reached a consensus on the equations for the new inflation, and cosmology suddenly had a new standard model: not just Big Bang, but Big Bang plus inflation.

Consensus, however, does not a science make. The work would continue — the tweaking and rethinking that Feynman's kind of faith eventually required. The whole field would be doing that tweaking and rethinking. The difference for Turner and Kolb was that they would be doing it Schramm-style — and with Schramm substance.

In the summer of 1981, during a hike in the Dolomites, Schramm and Leon Lederman, the director of Fermilab, discussed the idea of founding an institute devoted to the scientific intersection that Schramm had been championing for the previous decade: particle physics and cosmology. The idea was somewhat radical; as Turner said, "The two disciplines had little in common, other than indifference for one another." But NASA (perhaps as a consolation prize for awarding the Space Telescope Science Institute to Johns Hopkins University rather than Fermilab) agreed to fund it, and Lederman and Schramm hired Turner and Kolb to run the NASA/Fermilab Astrophysics Center.

"The Big Bang," Schramm often said, quoting Yakov Zel'dovich, a Russian theorist, "is the poor man's particle accelerator." Accelerators on Earth could approach the energies of the earliest moments in the universe — the earlier the moment, the higher the energy — but they couldn't match the earliest, most energetic moments. Even if you wanted to reach a time and a level of energy that an accelerator could match, it wasn't the kind of instrument you could roll down the hall and borrow for the afternoon. What you could do instead was perform the calculations on how certain particles at certain temperatures would behave, and then see whether those calculations matched the observations of elements in the universe today.

The idea itself wasn't new. Gamow, Ralph Alpher, and Robert Herman had tried to perform those calculations in the late 1940s, working from the assumption of a Big Bang universe. Fred Hoyle had tried to perform those calculations too, working from the assumption of a Steady State universe. The Big Bangers' calculations could account for the approximately three-quarters hydrogen and one-quarter helium abundances in the universe but not the 1 percent heavier elements. The Steady Stater had the opposite problem—able to account for the creation of the heavier elements but not hydrogen and helium. The twin impasse did nothing for cosmology's reputation.

In 1957, however, the physicists Geoffrey and Margaret Burbidge, Willy Fowler, and Fred Hoyle collaborated on a 104-page tour de force in the journal *Reviews of Modern Physics* that did for the origin of elements what Darwin had done nearly a century earlier for the origin of species. $B^2FH$, as scientists came to refer to the four collaborators, had worked for eighteen months in a windowless room in the Kellogg Radiation Laboratory at Caltech, scribbling on a blackboard, taking Baade and Zwicky's 1930s studies of the inner workings of supernovae to their logical conclusion, figuring out how nuclear reactions in successive generations of stars had ripped apart the basic building blocks of matter and put them back together in new and more complex combinations. Just as Darwin explained how single-cell creatures could evolve into species upon species, $B^2FH$ explained how single-proton atoms could eventually form the elements in the periodic table. As they phrased their conclusion, echoing Darwin's last line in *On the Origin of Species,* "The elements have evolved, and are evolving."

Gamow, Alpher, and Herman hadn't needed to account for the heavier elements after all. The hydrogen and helium they *could* account for were plenty, and then $B^2FH$'s supernovae would take over and produce the heavier elements. Following the discovery of the cosmic microwave background and the resurgence of interest in the Big Bang, physicists including Fowler, Hoyle, and Wagoner got to work on refining the calculations. The difference now for Schramm, and for the Fermilab Astrophysics Center he was deeding to Turner and Kolb, was that he wanted to get the physicists of the very small and the physicists of the very big talking to one another as if they belonged to one discipline—in fact, to create that discipline.

At once, Turner and Kolb started organizing "Inner Space/Outer Space," a conference that would advance the Schramm vision as well as the Schramm aesthetic. They adopted a logo that showed bubble-chamber tracks superimposed on a photograph of a galaxy, and they plastered it on souvenir T-shirts. They organized a jog around the accelerator track, the two hundred participants lumbering like bison. They showed off the buffalo that roamed on the Fermilab prairie. They hosted a "Buffalo Class* (*or equivalent protein material)" picnic, and advertised it with posters promising a talk by "J. Fonda* (*or equivalent protein material)." In the subsequent publication of the proceedings, they reproduced that poster and other whimsical drawings by Turner. And they included an afterword that doubled as a manifesto.

"Cosmology in the 20th century," they wrote, "has been hampered by a lack of confidence on the part of cosmologists, often leading to missed opportunities." Einstein lacked the courage of his equations: He missed predicting the expanding universe. A later generation lacked the courage of Gamow's equations: They missed discovering the cosmic microwave background. Their generation, Kolb and Turner vowed, wouldn't make that mistake. "Whatever future cosmologists write about cosmology in the 1980s, we can be certain it will not be that the cosmologists of this era were afraid to take even their wildest ideas seriously," they concluded. "We remain ever optimistic!"

The proselytizing continued (Turner even referred to NFAC as "the 'mother church'"). In 1989, Kolb and Turner published *The Early Universe,* a volume that they hoped would do for particle astrophysics and cosmology what Peebles's book had done for physics and cosmology, and then some. "Perhaps future cosmologists will laugh at our naïveté," they wrote in the "Finale" of the book. "But, if they do, we can hope they will admire our courage and boldness in attacking problems once thought to be beyond the reach of human comprehension. The extent to which we shall be rewarded for our courage and boldness remains to be seen. These authors," they concluded, "remain ever optimistic!" For the publication of the paperback edition they included a fresh preface: "Despite being four years older," they wrote, "we are no less optimistic; we are, in fact, even more optimistic!" And in recognition of their increasingly iconic presence in the community, they signed the preface "Rocky and Mike."

Fine: Be loud. Shout out your ideas. Crazy is okay, and the crazier the better. Snatch the chalk from your colleague. Standing to one side, leaning against a bar in a panelled basement, Schramm presumably appreciated the intellectual food fights on those Primordial Pizza nights.

But then: Be quiet. Cool off. Go back to your office in the morning and take out your notebook and refine the crazy idea until you've found the immovable maths. And make sure that that maths makes a prediction that someone can actually go out and verify. "Schramm's razor," his colleagues called this insistence on a verifiable prediction. And now cosmology had a prediction: The universe was flat.

So where was the evidence?

The "Inner Space" had changed over the years. In Schramm's original vision, inner space referred to particle physics, and he and his colleagues had succeeded in beating down the processes of element formation to what they called the "era of nucleosynthesis"—the period when the universe was between 1 second and 100 seconds old and the cosmic fog had cooled enough to allow the formation of elements. They knew what should have been happening in the previous fraction of a second, when protons and neutrons and electrons were ricocheting. But Hawking and Guth had changed the game; they came at the universe from the other end—not from the present backwards but the beginning forwards. They took into account not only particle physics but quantum physics. If inflation was right, then the quantum jiggling during the inflationary period—all $10^{-35}$ seconds of it—had frozen into the fissures in the cosmic pond, the veins in the ice, creating the structure around which matter (dark or not) had clustered, leading to the universe we see today.

But that "Outer Space" had been changing too. Long gone was the era of arguing over the Rubin-Ford effect—the data that Vera Rubin and W. Kent Ford had collected in the mid-1970s that seemed to show local galaxies not just receding in the expansion but moving en masse in a common direction. In 1976, the same year that Rubin and colleagues published the paper on the Rubin-Ford effect, a team led by Richard Muller and George Smoot at LBNL had taken a suggestion by Peebles in *Physical Cosmology* and, planting a Dicke radiometer aboard a U-2 plane, tried to measure the motion of our galaxy against the cosmic microwave background to determine whether the universe as a whole rotates. What they discovered instead was that our galaxy seemed to be racing through space at nearly 650 kilometres per second. Smoot made the announcement in April 1977 at an American Physical Society meeting during time that Peebles had yielded to him from his own talk. The phenomenon "is a real dilemma for theorists," Peebles said, and Smoot suspected that the two of them were the only physicists in the room who understood the implications: For the universe to contain such local volatility yet still appear homogeneous and isotropic on a large scale, that scale was going to have to be much larger than anyone had ever imagined.

That same year, Jim Peebles compiled a map of the millions of galaxies that the Lick Observatory had observed and found that not only did galaxies seem to be doing what galaxies interacting gravitationally with one another would be doing—clustering—but the clusters seemed to be doing what clusters interacting gravitationally with one another would be doing—superclustering. In 1981 Allan Sandage and Gustav Tammann announced that Gérard de Vaucouleurs (and, by extension, Rubin) had been correct: The Milky Way itself belonged to a local supercluster—indeed, the Local Supercluster, as de Vaucouleurs named it. That same year, a group including Robert Kirshner discovered evidence of the residue that clumping galaxies left behind: a "Great Void." The following year, another collaboration found that the Great Void wasn't so great; it was rather typical: "the superclustering phenomenon is widespread and accompanied by large holes in space that appear to be quite deficient of galaxies." A broader and deeper survey of galaxies by a Harvard Center for Astrophysics collaboration electrified astronomy by identifying a supercluster "Great Wall"—a filament of galaxies. But the Great Wall, too, in time came to seem typical, as redshift surveys continued to expand their reach. The pattern was consistent: The broader the slice of space, the longer the filaments; the longer the filaments, the greater the voids.

The scale itself of such structures presented a challenge for dark-matter theorists. Their simulations and calculations could show galaxies and clusters forming in the proper proportion in the distant past, but then the superclusters wouldn't have had time to develop later to the extent that observers were finding. Or their models could show superclusters having formed in the proper proportion in the recent past, but then the galaxies and clusters would have had to develop earlier to a greater extent than observers were finding. Still, for Turner and Kolb's purposes, at least the distribution of galaxies was, as one paper from this period reported, "frothy."

Did the galactic froth of Outer Space match the quantum jiggles of Inner Space? Since the discovery of the cosmic microwave background in 1965, the fate of the Big Bang theory had hung on the future detection of anomalies in the otherwise smooth bath of radiation — the inhomogeneities that had to be there in order for us to be here.

Was the universe flat? Since the invention of inflation in 1979, the fate of the quantum interpretation of those inhomogeneities had hung on the future detection of flatness.

In the early 1990s, Turner and Kolb and every other cosmologist who had been waiting for years got their answers to these two questions — to some extent. The Cosmic Background Explorer (COBE) satellite, launched in 1989, had been designed to make those two measurements at an unprecedented level of sensitivity — a level so sensitive that many scientists (including Turner) doubted that the experiment could work.

In 1990, John Mather announced that COBE had measured the spectrum of the microwave background and found it consistent with Penzias and Wilson's detection more than two decades earlier, and refined the measurement of the temperature to 2.735 K (plus or minus 0.06 K). In 1992, Smoot* announced that COBE had detected the wrinkles in the radiation in a proportion to match the predictions of inflation. The universe was flat.

Or not. At Princeton, for instance, Ruth Daly was using radio galaxies — galaxies spitting out plumes of plasma to either side so that they look like barbells — as standard measures. Like astronomers using supernovae as standard candles and hoping to see supernovae brighter (and therefore nearer) than they "should" be at cosmological

---

* Who by now realized that inflation would explain why his 1977 U-2 experiment might have failed to find the rotation of the universe: The universe as a whole might indeed be rotating, but we wouldn't be able to detect the effect in our little inflationary bubble.

distances, Daly and some other astronomers were hoping to see radio galaxies longer (and therefore also nearer) than they "should" be. Her preliminary observations corresponded to an omega of 0.1—one-tenth the density necessary to close the universe. Also at Princeton, Neta Bahcall was studying clusters of galaxies, hoping to extrapolate from their masses and distribution to "weigh" the universe. Her preliminary observations corresponded to an omega of 0.2—one-fifth the density necessary to close the universe. The universe according to Daly and Bahcall was open.

The universe was flat. The universe was open.

And that's where cosmology rested as the decade stretched on: a neither-nor state of suspension that would have to await further observations, a Pinteresque pause of cosmic proportions.

Late in 1997, this impasse assumed an importance beyond the professional for David Schramm when he received an invitation to participate in an event at the Smithsonian Institution's National Museum of Natural History. It would take place in April 1998, and it would feature a "Great Debate" between Schramm and Jim Peebles on whether the universe was flat. Such debates don't necessarily depend on the convictions of the participants; they're more rhetorical in nature. Still, Schramm's competitive side meant that he didn't want to simply score points. He wanted to be right. He wanted to know the value of omega.

He wanted to know what Saul Perlmutter knew.

For several years, Perlmutter's group had been promising a precise measurement of omega using Type Ia supernovae as standard candles. They had published several papers with hints of results, and now word was coming out of California that they were grabbing supernovae by the dozens, that they had won time on the Hubble Space Telescope, that they might be on the verge of delivering a verdict.

"What's Saul getting?" Schramm would say, winging past Turner's office. Then, a few days later: "What's Saul getting?"

Turner told him they might all find out soon enough, at the AAS meeting in January. Perlmutter would be making a presentation. The other team would be making a presentation. All they could do now was go home for the holidays and wait.

Late in the evening of 19 December, Michael Turner got a call. It was Judith, Schramm's wife. She was at their home in Aspen, Colorado, waiting for her husband, who was supposed to be flying there in his Swearingen SW-3, but he was hours late, and now there was a report on the local news about a plane crash.

Turner said something reassuring, and he and Judy promised to stay in touch. Even after she called back later, Turner found it impossible to believe that David Schramm had not survived the plane crash, had not wrested his impressive frame from the wreckage, was not even now wandering a field in the snow and the cold and the dark.

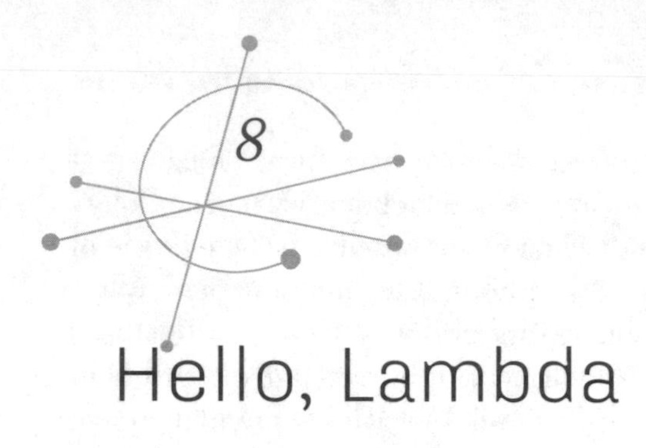


# Hello, Lambda

On 8 January 1998, four astronomers sat at a table at the front of a conference room at the Washington Hilton to deliver the verdict of science. Ruth Daly was there with her radio-galaxy data, and Neta Bahcall was there with her galaxy-cluster data, and representatives from the two supernova teams were there — Peter Garnavich for the High-z collaboration, and Saul Perlmutter for the SCP. The press releases from the various institutions had done their job. A couple of dozen journalists filled the seats, including reporters from the *New York Times*, and cameras on tripods lined the back of the room, their metal lamps throwing light and heat. The four astronomers represented four independent collaborations, but they spoke with one voice: The universe would expand forever.

One voice, however, was a little stronger than the rest. Perlmutter had flown to D.C. from an observing run in Hawaii. On the plane from Honolulu to San Francisco he had used a seatback phone for the first time, calling his colleagues in Berkeley and dictating the new data he'd collected in recent days at the Keck Telescope, atop Mauna Kea. Then he stopped in Berkeley just long enough to print out a poster incorporating that data. So far the SCP had made seven supernovae public, in a paper that had appeared in *Nature* a week earlier. But the team had more than forty other supernovae in the pipeline — a quantity that in itself was important. It communicated to the community that the system was working, and that the SCP had mastered it.

But for Perlmutter these results also represented the realization of his dream of using physics to solve the big mysteries. "For the first time," he announced at the AAS press conference, "we're going to actually have data, so that you will go to an experimentalist to find out what the cosmology of the universe is, not to a philosopher." Afterwards,he had stayed at a table in the room for an hour, conducting a mini-seminar for the members of the press. They surrounded him, and he held forth. Later, when they played back their cassette tapes, they might think they'd inadvertently hit FAST-FORWARD. But no, it was just Saul Perlmutter at regular speed, hyperkinetically trying to convince them that the headline here wasn't just the fate of the universe. It was that we could now *know* that fate—empirically, scientifically.

The following day Michael Turner paid Perlmutter a visit in the exhibit hall. The SCP team was part of the AAS meeting's poster sessions that day—dozens of presentations tacked to freestanding corkboards in long lanes, hard by the trade-show booths where representatives from weapons manufacturers sat at white-linen-covered tables and explained why the telescopes on their drawing boards were the best. Perlmutter wanted to show Turner something in the data, something he hadn't mentioned at the press conference.

Turner liked Perlmutter, and he liked the project; he didn't need to be convinced that the supernova survey was a worthy effort that deserved support from telescope time allocation committees or the National Science Foundation or the Department of Energy. Turner bent close to the panels—eight in all. The first few explained supernova search methodology to the uninitiated. One showed the logistics of the project: the initial observations at the Cerro Tololo 4-metre telescope, the follow-up observations at Cerro Tololo three weeks later, the spectroscopy at Keck, the photometry with telescopes at Kitt Peak and Isaac Newton and, at the highest redshifts, with the Hubble Space Telescope. The second showed light curves from twenty-one of the team's supernovae, the third showed some spectra, the fourth redshifts. The fifth explained how the SCP had calculated the photometry, made some corrections, and applied the stretch method to convert the Type Ia supernovae into calibrated candles, and how that calibration allowed them to plot the supernovae on a redshift-magnitude Hubble diagram. The sixth panel showed the Hubble diagram from the *Nature* paper, the basis for the claim that the universe will expand forever. Nothing Turner didn't already know.

But then came the seventh panel. It showed two plots. They were contour plots — plots that take the cumulative statistical effect (rather than individual points) of all the data and plop them on a graph that covers every possible scenario for the life of the universe. If your contour of confidence falls here, in this region, then you have a universe without a Big Bang. If it falls there, in that region, then you have a Big Bang universe that expands forever, and if it falls a little bit lower, then you have a Big Bang universe that eventually recollapses.

The plot on the left was, like the diagram in the sixth panel, from the *Nature* paper. It reflected the statistical effects of six supernovae, including the 1997 supernova that the SCP team had examined with the Hubble Space Telescope. And it did indicate that the addition of the one HST supernova shifted the likelihood up, into the region corresponding to a universe that expands forever.

The plot on the right, however, was new. It reflected the statistical effects of the dozens of other supernovae the team had found as well — forty in all. As you might expect, the addition of all that data had tightened the contours, narrowing the confidence regions. Looking at the graph on the left and then at the graph on the right was like putting on glasses; suddenly the fuzzy outlines of the world — the universe — snapped into focus.

The analysis was preliminary. But the effect so far was arresting. If you knew what you were seeing, you would get it at a glance. Yes, the universe was going to expand forever. But the evidence seemed to be indicating that even in order to exist, the universe couldn't be made up simply of matter, dark or otherwise. It needed something else.

Turner straightened up. "Dave would have liked that," he said.

Turner was at the AAS to lead a memorial service for David Schramm. Already he had attended one in Aspen, and he would be leading another later that month at Rockefeller Chapel, on the University of Chicago campus. Here were the supernovae that Schramm had been pestering Turner about, but if the data held up, here too was a hint of a further tragedy. Schramm had spent decades trying to rethink the universe, only to die, like Edwin Hubble at Palomar, in sight of the Promised Land — though perhaps even more poignantly.

In the half-century since first light, the telescope at Palomar hadn't delivered what Hubble had hoped it would: the two numbers that had kept his protégé Allan Sandage keening all the way into retirement. But if the SCP data held up, then science was entering an era Schramm had always envisioned: a new cosmology.

In 1917, in considering the implications of general relativity, Einstein saw that the universe was inherently unstable. Just as Newton had invoked God to keep his version of the universe from collapsing, so Einstein added a symbol to his equations — arbitrarily, the Greek letter lambda, Λ. Whatever lambda was, it was counteracting gravity, because, in Einstein's idea of a stable universe, *something* had to be. It was the reason that a universe full of matter attracting other matter through gravity wasn't collapsing. After Hubble's discovery of evidence for the expansion, the universe didn't need lambda, and Einstein had abandoned it. Unlike Newton's God, however, you couldn't altogether ignore it. Lambda was, after all, in the equation.

What you could do instead was set lambda to zero. That's what generations of observers and theorists had done. Sometimes they left the assumption implicit, simply failing to mention lambda. Often they stated the assumption explicitly: "Assume $\Lambda = 0$." For most observers and theorists, lambda was there and it wasn't there. It occupied a parallel existence, like a ghost in the attic.

Just because you didn't need it, however, didn't mean you couldn't invoke it, and from time to time theorists had done just that. In 1948, when Hermann Bondi and Thomas Gold and, separately, Fred Hoyle were trying to create a new model of the universe that didn't rely on an initial singularity of infinite density but still seemed to be expanding, they invoked what Bondi and Gold called the "hypothetical and much debated cosmological term." Like Einstein, they didn't know what it was, but they set it to non-zero because *something* had to be fueling the expansion. But then the validation of the Big Bang theory through the discovery of the cosmic background radiation eliminated the need for what had come to be called "the cosmological constant." Lambda didn't exactly die with the Steady State, but it fled the corpse, like a soul escaping.

It next took up residence in quasars, those mysterious sources of tremendous energy at mystifying distances. In 1967, a trio of Cornell theorists published a paper in the *Astrophysical Journal* examining, as the title said, "Quasi-Stellar Objects in Universes with Non-Zero Cosmological Constant." They were trying to resolve some possible inconsistencies in the behaviour of quasars.But as the understanding of the evolution of quasars became clearer, the need for lambda again receded. Then in 1975 two prominent astronomers argued in *Nature* that studies of elliptical galaxies as standard candles indicated that "the most plausible cosmological models have a positive cosmological constant." A year later they wrote another paper explaining why elliptical galaxies don't make good standard candles, implicitly undermining their earlier argument.

Four times now, including Einstein, cosmologists had gone up into the attic, and four times they'd returned with the same report: It was just the wind.

Then came inflation. It solved problems, the flatness and horizon problems. It explained improbabilities, the homogeneity and isotropy of the universe on the largest scales. And while the participants of the "Very Early Universe" Nuffield workshop at Cambridge in the summer of 1982 didn't agree on a model for inflation, they did, crucially, agree that a model could exist, and in the weeks and months after the workshop they formed a consensus around one model, giving inflation a solid basis in mathematics. But most important for its long-term survival or eventual obsolescence, inflation came with a prediction: that the universe was flat. That the amount of matter in the universe was equal to the critical amount that would keep it from collapsing. That omega equaled 1.

The problem for the inflation theorists, however, was that the observers were consistently finding evidence that the amount of matter in the universe was only 20 percent of the critical amount—that omega equaled 0.2.

At the final session of the Nuffield workshop, the theoretical physicist Frank Wilczek summarized the conference proceedings, concluding with "A Shopping List of Questions." Among them was whether omega was equal to 1. "If not," he said, "we must give up on inflation." Simple subtraction led you to conclude that for omega to equal 1 while observers were finding evidence that omega equaled 0.2, observers must be missing 0.8, or 80 percent, of the universe.

This discrepancy wasn't as worrisome as one might imagine. Two options immediately presented themselves. Maybe the rest of the matter was in a form that astronomers hadn't yet detected. The community had only recently conceded that the evidence for dark matter was compelling, and theorists were still working through the implications of dark matter for the structure and evolution of the universe. Or maybe the observers were just wrong, and more precise observations with improved instruments would boost omega and resolve the discrepancy.

A third option also existed, and if it, too, beckoned immediately, it did so from a distance, or even a different dimension. In any case, it was easy and probably advisable to ignore. Wilczek ended the Nuffield workshop with the last question of his "shopping list":

> What about the cosmological constant?
> "Whereof one cannot speak, thereof one must be silent."
> —Wittgenstein

Be silent? Be loud! Michael Turner went home from the Nuffield workshop, downed a slice of Primordial Pizza, and, along with fellow theorists Gary Steigman and Lawrence M. Krauss, got to work on a paper that explored the options for making omega equal to 1, entitled "Flatness of the Universe: Reconciling Theoretical Prejudices with Observational Data." Those "theoretical prejudices" referred to inflation's prediction of a flat universe, and the paper explored two ways of reconciling those prejudices with the data. One was a particle of some sort from the era of Big Bang nucleosynthesis—the field that Schramm had pioneered. The other possibility was "a relic cosmological constant."

"The cosmological constant," Turner liked to say, "is the last refuge of scoundrel cosmologists, beginning with Einstein." He himself, in his "heart of hearts," thought the cosmological constant must be zero. But he also knew that the cosmological constant had "every right to be there." And as he and Rocky Kolb often insisted, their generation wasn't going to make the mistake that Einstein and other twentieth-century cosmologists had made by not taking every remotely serious option seriously.

If anything, the self-described scientific conservative Jim Peebles took the idea even more seriously than Turner—but then, Peebles prided himself on trusting observations more than most theorists.

"What's best," he would say, shrugging, "is what's true." The truth for him had emerged in a 1983 paper he wrote with Marc Davis, a UC Berkeley astronomer, using the latest and largest survey of galaxies to measure their velocities, infer their masses, and derive the mass density of the universe. Peebles looked at their data and thought, "High mass density is dead in the water." Their conclusion: an omega of 0.2.

The following year, Peebles wrote a paper, "Tests of Cosmological Models Constrained by Inflation," that offered his theoretical interpretation of that data. Maybe omega was indeed 0.2 and lambda equaled 0, he wrote, but in that case "we lose the attractive inflationary explanation for the observed large-scale homogeneity of the universe." He didn't *want* a cosmological constant. "It's ugly," he often said. "It's an addition." If he were building a universe, he thought, he wouldn't put in a cosmological constant: "No bells and whistles." But perhaps because inflation solved the flatness problem he'd articulated with Dicke, or perhaps because he constitutionally distrusted a simple universe, he accepted the possibility with equanimity. "Considering the observations," he said, "I think the universe might have put in bells and whistles — a cosmological constant."

The paper met with a lot of resistance, which Peebles sort of enjoyed. He found that he could go to conferences and give a talk, and people would rant at him, and then they apparently forgot, because a few months later he would give the same talk and the same people would rant at him. He realized he didn't even have to write a new talk; he could just give the same one over and over. This went on for a decade.

Theorists are always saying something. That's their job. They don't need to believe what they're saying. The theorist's goal isn't to be right but to be reasonable — to make an internally consistent argument that observers can then go out and reinforce or disprove. For their part, observers regard theorists with patience and exasperation, like a dog that's always depositing gifts at their feet: a stick, a squeaky toy, a dead bird. Often these offerings just lie there. But once in a while the observers will throw them a bone. *Go fetch.*

In 1992 observers threw cosmological theorists the biggest bone since the discovery of the cosmic microwave background more than a quarter of a century earlier: the Cosmic Background Explorer results — the ones that said that the universe was flat. The following year, Turner and Kolb added a preface to the paperback edition of *The Early Universe* reviewing the COBE results and declaring them "a shot in the arm" for a flat universe.

As other observations accumulated that indicated a universe with a low density of matter — especially the kind of studies of galaxies on the largest scale that had first persuaded Jim Peebles a decade earlier — theorists found themselves increasingly less reluctant to suggest, and observers found themselves increasingly less reluctant to consider, the possibility of a cosmological constant. "WHY A COSMOLOGICAL CONSTANT SEEMS INEVITABLE" read a section heading in one influential paper; "The Observational Case for a Low-Density Universe with a Non-Zero Cosmological Constant" was the title of another paper. And then there was Turner again, again with Lawrence Krauss: "The Cosmological Constant Is Back." The cosmological constant was still the last refuge, but it was a refuge nonetheless.

Vera Rubin summarized the situation with a joke. There was a wise rabbi, she said, who was trying to mediate a marital dispute. The husband complained about the wife. "You're right," the rabbi said. The wife complained about the husband. "You're right," the rabbi said. Then the rabbi's own wife emerged from behind a curtain, where she had been eavesdropping. "How can you tell them both they're right?" she said to her husband. To which the wise rabbi replied, "You're right too."

She told this joke at a "Critical Dialogues in Cosmology" conference at Princeton, part of the university's celebration of its 250th anniversary, in the summer of 1996. The purpose of the conference was to bring together the world's leading cosmologists to address the field's greatest challenges. One such event, inevitably, involved the value of omega, and it took the form of a debate. On one side was Avishai Dekel, who had recently measured galaxy motions that were consistent with an omega equal to 1. On the other side was Turner, arguing that the amount of matter in the universe was not enough to nudge omega to 1. But he didn't stop there. Instead he used the forum to argue that omega was indeed equal to 1, because the cosmological constant would close the gap.

The moderator was none other than Bob Kirshner. At one point in the discussion he turned to Saul Perlmutter, who had arrived at

Princeton bearing preliminary results from the SCP's first seven supernovae. Kirshner asked what he thought.

Like any cosmologist dealing with omega, the SCP had addressed lambda in paper after paper: "(for $\Lambda = 0$)," "(for $\Lambda = 0$)," "If we assume that the cosmological constant = 0." A year earlier, in 1995, Perlmutter and Ariel Goobar had elevated the cosmological constant from its pro forma purgatory, making its existence the subject of a paper in the *Astrophysical Journal.* Or, rather, making its *non*existence the subject of the paper, since their assumption while writing it was that matter would indeed account for everything. They figured that they would be explaining how astronomers could use supernovae to show, once and for all, that lambda equals 0.

And that's what Perlmutter had come to Princeton ready to discuss. Yes, he reported. Now the SCP's first seven supernovae were consistent with a universe where omega equals 1 and lambda equals 0.

"This could be a lambda killer," Jim Peebles told a journalist.

*Lambda killer.* Perlmutter liked the sound of that: Get lambda out of the way so it won't be a spoiler anymore.

Like his mentor David Schramm, Michael Turner didn't like to lose debates. And like Schramm, he wasn't afraid to practice what the Fermilab and Chicago cosmologists called "jugular science." During a break in the Princeton activities, as various astronomers and cosmologists were climbing a flight of stairs to an auditorium, Turner sent Perlmutter a message. Ostensibly talking to the astronomer walking beside him, Turner raised his voice.

"I don't think Saul is that stupid," he said.

Perlmutter didn't appear to hear.

"I said," Turner repeated, raising his voice, "'I don't think *Saul* is that *stupid.*'"

Turner was slightly more diplomatic during his own talk, but no less needling. "I am anxiously awaiting the results of the two deep searches for supernovae," he said, referring to the rival teams. "I think they're going to shed some important light on this. To draw any conclusion now would be to take away from their thunder later."

The SCP submitted their data on their first seven supernovae to the *Astrophysical Journal* that August. If they made the standard assumption regarding lambda—"a $\Lambda = 0$ cosmology"—then omega was 0.88. But given the margin of error, you could reasonably interpret that result as omega equaling 1. If they made the less likely assumption that the universe was flat with a possible component of lambda, then omega, at 0.94, was even closer to 1, while lambda would be 0.06, a negligible amount and, given the margins of error, presumably 0.

The universe was flat. Matter alone was enough to get omega to 1. And we didn't need lambda. Or at least that interpretation was, as the paper said, "consistent" with their results.

Unfortunately for the SCP team, that interpretation wasn't consistent with their own next round of data.

The *Astrophysical Journal* accepted the paper in February 1997 and published it in the 10 July issue. By then the SCP team was finishing their analysis of the two supernovae they'd examined with HST. Because HST photometry would be so superior to ground-based analyses, the team would place special emphasis on whatever it had to tell them.

Peter Nugent had been hired as a postdoc by Perlmutter a year earlier, part of a campaign to bring astronomers into the project. Nugent had written his thesis on Type Ia supernovae, and Perlmutter had assigned him to perform photometry. Nugent had a forceful style. He wouldn't have been out of place at the University of Chicago; his bearing and attitude were reminiscent of a David Schramm or a Rocky Kolb: a can-do, answer-any-question, know-the-restaurants-with-the-best-wine-lists spirit. On 30 June he finished the photometry on the two HST supernovae, giving him their magnitudes, the standard measure of luminosity for celestial objects. Spectroscopic analysis had already yielded the redshifts for the two supernovae. Now Nugent plotted the two values against each other, redshift on one axis, magnitude on the other.

You would expect the points on this plot to land pretty much on the usual 45-degree-angle straight line — the relationship among nearby galaxies that Hubble discovered in 1929. The straight line itself represents a universe that is expanding uniformly, experiencing no effects of gravity — in other words, a universe without mass, a universe with nothing in it. Eventually, at some great distance across space and back in time, the points will have to begin to deviate from the straight line to represent a universe that does have mass. But which kind of universe? The extent to which the most distant points deviate downward from the straight line will be minor, but it will tell you how much brighter the objects are than you would expect them to be at their particular redshifts — the brighter the supernovae, the higher the value of omega. And that value will tell you the weight, shape, and fate of the universe: open, closed, or flat; saddle, globe, or plane; Big Chill, Big Crunch, or Goldilocks.

Nugent began plotting the two HST supernovae. First he looked along the redshift axis — the measurement that corresponded to their distances. Then he moved up the graph until he reached the magnitudes his photometry had given him. He assumed the two points would land along the deviation — the particular downward curve — consistent with the conclusion that the SCP's latest paper had reached: a flat, all-matter, omega-equals-1 universe. But that's not where these two supernovae fell. They were landing on the other side of the straight 45-degree-angle Hubble relationship, on what would be an *upward* curve. The difference between what their luminosities should be at their redshifts and what their luminosities were, was approximately half a magnitude, meaning that the two supernovae were 1.6 times *fainter* than he expected.

"There goes the universe," he wrote in the e-mail to his team. Not that he was ready to draw any conclusions about cosmology. After all, as he wrote, "it's only two data points." And he wasn't the team member responsible for determining the omega and lambda measurements. But the discrepancy between the magnitudes he expected and the magnitudes he measured was unequivocally jarring. "Hopefully this will be enough from me to get the paper out this week," he added. "I do think it has to go out now since the other group is most likely going to submit something soon (very soon) [about their own HST results]. It's good with the data — as-is. Lets get the damn thing out there!"

But they didn't. The team quickly realized that they needed to decide not only whether to publish, but what.

In the jargon of science, the two HST supernovae were "fighting" the earlier, all-matter, omega-equals-1 result. That summer, the team threw out two of the first seven supernovae — one that further analysis determined to be a core-collapse supernova rather than a Type Ia, and another one that was an obvious outlier. They also eliminated the 1996 HST supernova because they felt that, while the individual measurements were probably accurate, they didn't have enough observations—enough points on the light curve—to subject the supernova to peer review. But the similarity in results between the 1996 and 1997 HST supernovae did reinforce the team's confidence in the 1997 data. By September they had settled on six supernovae in total as well as a conclusion—albeit one inconsistent with a paper they had published only two months earlier.

One member of the collaboration wrote to Nugent that he "must realize that we will look very bad if we change our limits every time we add *one* SN to the total sample without a discussion [in the paper]. How can anybody trust what we say if they know we are going to say something else in a few months time without any explanation?" After all, the two papers weren't dependent on two separate samples or a significantly larger set of data. The new paper had two *fewer* supernovae than the previous paper. The only addition to the data was a single supernova from HST.

The point of the paper, Nugent argued back, would be to demonstrate what HST could do for a distant supernova search—"NOT" to declare that the universe has certain values of omega and lambda, "says God." The number of supernovae didn't matter. "I've never given a rat's ass about one data point (or even a number under 10 for that matter) in my life when the error bars are so large." The *method* was what was worth reporting.

Still, writing a paper that reverses a result even implicitly was going to require some finesse. Not until September 1997 did the team have a draft they would submit to *Nature*, and by then they had larded the prose with enough qualifiers to choke even a Kirshner: "we use the words 'preliminary', 'initial' and 'if . . .' all over this paper," Nugent reassured a colleague in a 27 September e-mail. And when the paper got to the omega and lambda part, it delivered a double qualifier: "these new measurements *suggest* that we *may* live in a low mass-density universe" (emphases added).

The team submitted its paper on the HST supernova to *Nature* the first week of October. Sure enough, just as Nugent had fretted at the end of June, the High-z team followed with its own HST supernova paper, with Garnavich as the lead author, posting it on the Internet on 13 October. The High-z paper reported that its sample, too, "suggests that matter alone is insufficient to produce a flat Universe." Clearly the two groups were converging on the result that had motivated their supernova searches: the fate of the universe.

If there was no cosmological constant, then omega was low and the universe was open—destined to keep expanding for all time. Even if there was a cosmological constant, then omega was still low and the universe was flat—slowing to a virtual halt, but not collapsing. Either way, the expansion of the universe would continue forever. That autumn the American Astronomical Society invited both teams to participate in a press conference at the AAS meeting in January 1998. The PR department at the AAS usually organized four or five press conferences during the course of the semiannual five-day meetings and a discussion of the fate of the universe seemed like the kind of topic that would draw a crowd. Sure, the two teams told the AAS, we'd be glad to send representatives to a press conference.

But a subtler, and certainly more esoteric, question remained: *Was* there a cosmological constant?

Gerson Goldhaber, anyway, thought there was. On 24 September he showed the group the histograms compiling all the supernovae, one for a no-lambda universe, and one for an omega-plus-lambda-equals-1 flat universe. For a measurement as delicate as the one the team was trying to make, binning supernovae into broad categories wasn't going to be as persuasive as plotting individual points. But a trend was clearly developing. The more supernovae the team analysed, the lower the value of omega seemed to be heading. Two weeks later, the minutes from another team meeting reflected the trend: "Perhaps the most disturbing thing is that the first 7"—the bunch on which the team had based their previous paper—"were consistent within themselves but the next 31 Sne give what seems to be a consistent answer that is lower."

In the 1930s Fritz Zwicky had discovered a set of supernovae that he assumed were examples of the *implosion* process that he and Walter Baade had predicted; in retrospect, those supernovae all turned out to be examples of an *explosion* process that hadn't yet been discovered. Now the SCP team was realizing that they, too, had defied the odds. Even after eliminating the obvious outlier and the Type II from the original set of seven, those five initial supernovae still appeared to be on the bright side. As a result, the addition of the dozens of fainter supernovae was driving the value of omega down. In the histogram analysis of the data, a sharp peak was developing around an omega of 0.2.

On 14 December 1997, Goldhaber presented his findings at a seminar at the Institute for Theoretical Physics at UC Santa Barbara. Kirshner was in residence at the institute that autumn, on sabbatical from Harvard, and as usual Goldhaber found him to be "antagonistic." Kirshner interrupted the presentation: An omega of 0.2; so what else is new? But Goldhaber thought he was making an argument that omega could be 0.2 only if accompanied by lambda. At least the director of the institute, David Gross, seemed to understand, though when he asked Goldhaber why he believed the results, all Goldhaber could offer was that he had a long history of interpreting histograms. "I'm convinced," he said.

Perlmutter, too, was presenting preliminary results in public that term, carrying his transparencies of low-omega scatter plots from colloquium to colloquium—the first on 23 October at the Physics Department at UC San Diego, the second on 1 December at the Physics Department at UC Berkeley, and a third on 11 December at the Physics Department at UC Santa Cruz. As in the *Nature* paper, he was careful to qualify his comments, but he also made sure to let his audiences know that the data contained the possibility of "some rather striking consequences for physics," as he said at the Berkeley colloquium. "In particular, if you consider the flat-universe case—the case of the inflationary universe that's favored—a mass of this sort, a mass density of this sort, means that the cosmological constant has to be contributing a cosmological constant's energy density of about 0.7." In case the non-cosmologists in the audience were missing the point, the astrophysicist Joel Primack stood up at the end of Perlmutter's talk at Santa Cruz to say the results were "earthshaking." Then he added the crucial caveat: "If true."

For the High-z team, Adam Riess was now "it." Riess knew that his team was at a disadvantage concerning the quantity of supernovae, if only because Peter Nugent kept reminding him. The two of them were in a group that got together on weekends in a city park to play a variation on American football called, for obvious reasons, mudball. Sometimes the banter took the form of my-distant-supernova-search-is-better-than-yours. One day Riess decided he was tired of hearing how many supernovae the SCP was raking in and how far behind the High-z search was. If you couldn't beat the SCP on quantity, he figured, you could beat them on quality.

For his master's thesis Riess had tackled the problem of dimness. His light-curve shape method proposed a mathematical solution to deriving luminosity from the rise and fall of light-curve shapes. For his PhD thesis Riess had approached the problem of dust. If you're trying to determine the distance of a supernova by measuring its red-shift, then you need to know to what extent dust is contributing to the reddening of the light (just as dust in the atmosphere reddens a sun-set). In Riess's multicolor light-curve shape method, or MLCS, the observations of light in several color filters would provide a cumulative measure of the effect of dust, allowing you to derive a more accurate determination of distance.

As his team's resident expert on correcting for intergalactic dust between the supernova and the observer, he might be able to clean up the supernovae in such a way that they provided a tighter margin of error than the SCP's. He wouldn't even need a greater number of distant supernovae, though they were always welcome. Even nearby supernovae would do the trick. If he could anchor the lower end of the Hubble diagram with sufficiently reliable data, then the higher-redshift supernovae — while fewer in number than the SCP's — would be more reliable as well. And he knew where he could get nearby supernovae: observations he had already made, as part of his thesis research, at the 1.2-metre telescope on Mount Hopkins, in Arizona. Twenty-two supernovae in all. None of them yet published.

The addition of those supernovae, however, created a new problem. Never mind a universe with no matter. His calculations were producing a universe with *negative* matter.

"I'm only a postdoc," Riess told himself. "I'm sure I've screwed up in ten different ways." Computers, he thought, don't know physics. They know only what we program them to know. Clearly he had programmed his computer with impossible physics. So Riess checked his maths, and he checked the computer code he'd written, and he couldn't find any mistakes. Of course, Einstein's equations allowed for another option—a universe with a positive lambda. Plugging his data into *that* universe brought the amount of matter up, into the positive range. But that option, he knew, wasn't palatable to most astronomers—for instance, the team leader, Brian Schmidt, who liked to say that astronomers who talk about the cosmological constant are astronomers without many friends.

Riess sent his results to Schmidt.

"Adam is sloppy," Schmidt reminded himself. Brilliant, but prone to mathematical errors. Schmidt agreed to double-check the results. As a rule, mathematicians check each other's work not by looking back over the same calculations but by performing the calculations independently, so as not to be lulled into making the same mistakes. Schmidt and Riess soon developed a routine. Riess would e-mail a problem, and a day later Schmidt would respond. *I started with this image, and my analysis said the supernova was this bright—how about you?* Or *We observed in this filter, and I found that the redshift was equivalent to this number—how about you?* They signed their e-mails Pons and Fleischmann, after the two physicists who, in 1989, had "discovered" cold fusion, and who, after a long period of infamy, had fallen into obscurity. If you're Stephen Hawking and you make a major mistake, you're still Stephen Hawking. If you're a postdoc under thirty and you make a major mistake, you're history. Sometimes when Riess couldn't wait for an answer, the phone would ring in the Schmidt household. Schmidt's wife, sleepless from caring for a six-month-old, would say, "If that's Adam, tell him to—"

Schmidt: "Hello, *Adam.*"

Riess: "Oh." A pause. "Is it early there?"

Schmidt: "It's four in the morning."

Riess: "Oh." A pause. "So, what do you know?"

What Schmidt knew, night after night, was the same thing: So far, so good.

Riess remembered now that one day when he was a graduate student at Harvard, Kirshner had brought Mike Turner and Alan Guth by his office and encouraged Riess to show them what he was working on. Riess had just taken the team's first Type Ia supernova, 1995K, and plotted it on the Hubble diagram. The supernova fell on the "bright" side of the 45-degree straight line, but its location didn't matter; it was only one point. What mattered was that the team actually had a point to plot. Still, Turner couldn't help mentioning that it was in the "wrong" part of the diagram.

"How embarrassing," Riess had thought. "There's never been so much brainpower in P-306"—his office at the time—"and we're probably showing them that we're not even doing the experiment right."

But now, a couple of years later, he thought maybe the location of that point had mattered more than anyone knew. Maybe the answer to the fate of the universe had been right in front of them from the very first supernova.

Riess was getting married in January 1998—the weekend at the end of the AAS meeting, in fact. When his future wife flew home from Berkeley to her family in Connecticut a few weeks early to take care of the final preparations, Riess sequestered himself in his office in Campbell Hall, on the Berkeley campus, and began writing a paper that would report the results—if they held up.

The campus was empty for holiday break. The heating was off, and Riess had to bundle up; even in California, December can get chilly. But every day, walking past the locked office doors and under the unlit hallway lights, he went to work. On 22 December, he wrote an outline and started a draft. Garnavich's HST paper, the first from the group, had been a short letter. Riess figured the next paper would have to be the *War and Peace* version, as scientists like to say; if you're claiming something surprising, you have to show all the work. In the coming days he also contacted Nick Suntzeff, down in Chile, and asked him to double-check some photometry, though he didn't say why so as not to prejudice the result. At one point he beckoned a colleague into his office.

Alex Filippenko, who also had been taking advantage of the semester break to catch up on work, greeted Riess with his usual wide and deep smile, rectangular and cavernous. Nobody could be that happy all the time, and Filippenko wasn't. He had once been a member of the SCP, and as an astronomer on a team with a particle physics mentality, he had experienced the clash of cultures probably more acutely than anyone else on either team. He disliked the hierarchical structure that awarded Perlmutter lead authorship on the important papers; Filippenko would go to astronomy conferences and hear about "this supernova survey" that Perlmutter had organized, and he'd have to inform his peers that he was actually part of that collaboration. watched as his friends in the supernova game—Kirshner, Riess, Schmidt, Suntzeff—coalesced into a collaboration of their own. He complained to them that he had been warning his SCP colleagues about the possible non-standardness of Type Ia, about dust, about the difficulty of photometry and spectroscopy—all the concerns that Kirshner had been raising for years as a member of the External Advisory Board. He said he felt that the Berkeley Lab physicists regarded these concerns as if they were "irritations" and "annoyances" rather than supernova astronomy's swords of Damocles. He felt marginalized and ignored on the SCP collaboration, and he suspected that they kept him around only as the "token astronomer" who could get them time on telescopes.

But all his friends could do was shrug and say that yeah, they would love him to be on their team, but he was part of the other team.

In early 1996, Filippenko defected. A few months later he was able to exact a revenge of sorts. As Riess was finishing his PhD work at Harvard, Perlmutter approached him with the offer of a position at LBL and, by extension, on the SCP team. Down the hill at UC Berkeley, Filippenko countered with a Miller Fellowship—the same honor that Filippenko himself had once held as a young postdoc.

Riess didn't have to think too hard. He'd be doing astronomy with a friend. He'd be doing astronomy on a team to which he already belonged. He'd be doing *astronomy*.

Exactly. For Filippenko, Riess's MLCS method was precisely the kind of tool that an astronomer would know was necessary—would feel a need to invent—before proceeding to the next step. And now Riess apparently wanted to show him where that next step had led.

Riess pointed to the notebook on his desk. He walked Filippenko through the calculations he'd made, he described the back-and-forth that he and Schmidt had been having, and he said that the result didn't seem to be going away. Filippenko studied the notebook for a few moments, then straightened up. He was shaking his head.

"Man," Filippenko said, "be sure the measurements are done right."

Well, of course.

By 4 January, Riess had taken the paper as far as he could. He sent Schmidt the material for the final round of cross-checking. Then he waited.

"Well Hello Lambda!" Schmidt e-mailed him on 8 January, the day of the AAS press conference. Schmidt had finished his spot checks and found nothing wrong. His statistical level of confidence was the same as Riess's: 99.7 percent. It was time to let the rest of the team know.

When Pete Garnavich got to the AAS meeting, he had already studied the SCP's paper in *Nature,* and as the lead author on the High-z team's HST paper—due to appear in the *Astrophysical Journal* on 1 February—he certainly would have sought out the SCP's even more recent data on his own. But he knew, too, that his team was getting a bizarre result; Riess and Filippenko had confided it to him before the press conference. They had also instructed him to keep the result quiet. On 9 January, Garnavich visited the AAS poster session meeting to see for himself how close the SCP was to getting that same bizarre result.

Close.

Clearly the SCP supernovae were falling in the low-omega range. The fate of the universe was to expand forever. Ho hum.

But what about a cosmological constant? Could the SCP claim evidence for a non-zero lambda? Could they say with some conviction that the universe would have negative mass—essentially, that it wouldn't exist—without the addition of a positive lambda to the equations?

Not quite, as far as Garnavich could see. The error bars above and below the points on the graph representing the supernovae could certainly accommodate such an argument. Some of the upper limits, and some of the supernovae themselves, fell within the range of the upward curve designating a universe with non-zero lambda. But some didn't. Garnavich concluded that the SCP wasn't ready to claim anything explicit, anything definitive. He reported back to his colleagues that High-z was still in the lambda game.

Riess flew east for his wedding on 10 January. He returned to Berkeley two days later, for a one-night stopover on his way to his honeymoon. That evening he checked his e-mail. The string of boldface unread messages stretched down the screen. Riess scrolled. Still it stretched. When he got to the bottom of the list, he checked the time stamp to see how long the conversation had been going on without him. Forty-eight hours — "an eternity."

He started at the bottom — a question from Schmidt: "how confident are we in this result?"

"In your heart," Kirshner wrote back, "you know this is wrong, though your head tells you that you don't care and you're just reporting the observations."

"I don't know about anyone else," another team member responded later that day, "but MY heart tells me nothing about the cosmological constant." They had a result; they had a confidence level. Which led to the second question: Did they need to believe it in order to publish it?

Kirshner didn't want to risk reporting evidence of a cosmological constant that they would have to retract later. "That would be like saying 'Omega must be 1' based on 4 supernovae and then saying 'Omega must be Zero' when you get one more. Perlmutter has already done that. He's a year ahead of us, but I don't think we want to duplicate that path!"

Bruno Leibundgut, writing back from Germany that same day, agreed. "There is no point in writing an article if we are not very sure we are getting the right answer."

Mark Phillips, in Chile, concurred. "Press releases and a barrage of ApJ Letter/Nature articles may impress the public or scientists who have only a casual interest in the subject, but the hard-core cosmology community is not going to accept these results unless, as Bruno says, we can truly defend them."

Schmidt, however, disagreed. "As uncomfortable as I am with a Cosmological constant," he wrote, "I do not believe we should sit on our results until we can find a reason for them being wrong (that too is not a correct way to do science)."

Correct way or not, there was a further concern: priority.

"Of course we want to remain true to our scientific ideals," wrote another member of the collaboration. "But this has to be balanced with realpolitik."

"Who knows?" Filippenko wrote. "This might be the right answer. And I would hate to see the other group publish it first." And if it's the wrong answer? Another team member argued that there was no downside: "If it turns out in the fullness of time that a cosmological constant exists they" — the SCP — "can claim to have found it. If it does not, their claim will be forgotten and no one will attach much blame to them for being wrong." Reporting that same conclusion before SCP did was was a no-lose proposition. If the High-z team was right, they would get priority; if they were wrong, they'd get a free pass. High-z had a reasonable argument, so why not make it?

Why not? Riess saw no reason. He leaned into his keyboard and started to compose a response that he could send to the whole group. When he looked up from the screen, he found his bride staring back at him.

"I cannot believe," she said, "you are working on an e-mail when we are on our way from our wedding to our honeymoon."

"Well, this is a really important one."

"Oh, I think I'm going to be hearing this all the time."

"No, no," he said. "This — you're not. This really is — this is the one."

She shook her head and left the room. Riess bent back to the keyboard, composing an e-mail that would answer all the questions.

Heart or head?

"The data require a nonzero cosmological constant!" he typed. "Approach these results not with your heart or head but with your eyes. We are observers after all!"

Publicity? Priority? Realpolitik?

"You see, I feel like the tortoise racing the hare. Everyday I see the LBL guys running around but I think if I keep quiet I can sneak up. . . . shhhh . . ."

Finally, quick letter or *War and Peace*?

"I think I can answer the group's dilemma about a quick kill paper vs a detailed explanation. . . . you all said you wanted a detailed exposition of the data so that's is what I have been working on. Brian said that adding the data stuff to the paper should only take a week, well I did it already before the wedding."

He hit SEND, and the next morning he left for his honeymoon in Hawaii. (Also, an observing run at Keck.)

For Perlmutter, the extra effort he'd put into the preparations for the AAS meeting had paid off. The media coverage focused primarily, and rightly, on the consensus that the participants in the press conference had reached — the fate of the universe. The *New York Times* ran it on the front page, under the headline "New Data Suggest Universe Will Expand Forever." The *San Francisco Chronicle,* the hometown paper for the SCP team, had also put the news on page 1. The local paper for the AAS meeting, the *Washington Post,* ran its story on page A3: "Universe Will Keep Expanding Forever, Research Teams Say." But it was the SCP that the *Post* singled out for a rave. "Perlmutter bowled over the audience with an unexpectedly large sample," the article said. "Garnavich's team presented three." And then, of even greater significance to the astronomy community, came a news article in the journal *Science* three weeks later.

The author of the article, James Glanz, a PhD in astrophysical sciences, covered his beat as if it were City Hall. He had written about the supernova searches over the past several years, but his most recent reporting mentioned a possibly imminent discovery. In the 31 October issue he wrote that both teams had submitted papers supporting the conclusion that the universe would expand forever — a scenario that held whether the universe was open or flat, whether omega was less than 1 or exactly 1. But then he added that such a never-ending expansion would be "perhaps boosted by large-scale repulsive forces."

"The results," he continued, a few paragraphs into the article, "still leave an opening for some theories in which matter plus its equivalent in energy, supplied by the cosmological constant, add up to a flat universe."

The article in late January also included a reference to "a quantum-mechanical shimmer in empty space, called the cosmological constant," but this time Glanz focused on the SCP's contribution to the AAS meeting:

> Not only did the results support the earlier evidence that the expansion rate has slowed too little for gravity ever to bring it to a stop; they also hinted that something is nudging the expansion along. If they hold up, says Perlmutter, "that would introduce important evidence that there is a cosmological constant."
>
> "It would be a magical discovery," adds Michael Turner. . . .

Since the frantic exchange of e-mails early in the month, the High-z collaboration had been trading drafts, exploring the maths, debugging the code (Schmidt and Riess had missed a few glitches, but nothing important), examining the photometry and the spectroscopy and the charts and graphs and tables—all in the cause of making their case scientifically responsible. Now they had a further concern. Glanz's article, complete with a reproduction of a contour graph showing SCP's preliminary analysis of forty supernovae, seemed to be suggesting that the SCP was beating them at beating the SCP at beating them at their own game.

Alex Filippenko would be speaking at the upcoming UCLA Third International Symposium on Sources and Detection of Dark Matter in the Universe, in Marina del Rey. The High-z team would be submitting their paper only a couple of weeks after that. Why wait? he asked. Filippenko suggested he could announce the team's findings at the UCLA meeting. "This is our chance to make a big splash," he said.

But why not wait? argued other members of the team. The paper will be out soon enough. Let the science speak for itself.

Filippenko, however, retorted that if the SCP was as close to claiming a discovery as their AAS presentations suggested, then those two weeks might make a crucial difference in terms of establishing priority. "You can check and recheck your results forever," he said, "but at some point you've got to have the balls, basically, you've got to have the courage, to announce your result and to say, 'Okay, here is an accounting of our uncertainties. This is where we stand.'"

The discussion stalled there. But then, just days before the UCLA meeting, Jim Glanz called Filippenko. He didn't know whether Glanz was using the old reporter's trick of pretending to know more than he did, but for Filippenko the conversation was all he needed to convince a majority of the High-z team. "Glanz is going to be breaking this story whether we're in it or not," Filippenko said. "So why not be in it?" Give him the evidence. Give him the quotes. And give him the news peg.

Make the announcement, and tell him about it first.

On 22 February, Filippenko took his seat at the UCLA conference and listened as Gerson Goldhaber gave a presentation on the SCP team's latest results. Then he listened as Saul Perlmutter gave a presentation on the SCP team's latest results. As far as Filippenko could tell, nobody was claiming a discovery; all he heard was that the SCP had "evidence" for lambda.

He took a deep breath. It was now his turn to present. Filippenko stood up, paused, and then said either you had a result or you didn't. And the High-z team did.

The ghost was real, and it was most of the universe.

# The Tooth Fairy Twice

MIKE TURNER WAS following in David Schramm's footsteps. He was walking along the hallways and footpaths, among the blackboards and picnic benches, of the Aspen Center for Physics, a summer retreat for theorists under head-clearing blue skies. One look at the mountains, one deep breath, and you could see why a big-as-all-outdoors guy like Schramm had fallen in love with the place at first sight in 1976, enough to make Aspen his second home. Eventually he'd served as the chairman of the board for the Aspen Center, from 1992 until shortly before his death. But Schramm was gone now, and Turner had agreed to take his place opposite Jim Peebles in the "Nature of the Universe Debate" at the Smithsonian, so when Turner ran into Peebles at the Aspen Center, he had a question for him. For obvious logistical reasons, the organizers had bumped the event from April 1998 to October—and just as well. Turner and Peebles needed a new topic.

"Are you still willing to debate non-flat?" Turner asked.

Peebles shrugged.

"Jim, debates have a yes-or-no question. Correct me if I'm wrong, but you and Dave were supposed to debate whether or not the universe is flat. He got flat and you got non-flat." Turner asked his question again. Did Peebles really want to argue publicly that the universe wasn't flat?

"No."

The answer didn't surprise Turner. Both theorists knew that defending a non-flat universe in late 1998 would feel like defending the

Steady State cosmology in late 1965. Within months of the January AAS press conference where Perlmutter unveiled the SCP's forty-two supernovae, and within weeks of the February UCLA meeting where Filippenko made his announcement, a consensus had emerged in the Big Bang community of astronomers, astrophysicists, cosmologists, and theorists: The universe wasn't what it used to be.

Since Hubble's discovery of evidence for the distance–velocity relation, astronomers had been following a syllogism: One, the universe is expanding; two, the universe is full of matter attracting other matter through gravity; therefore, the density of matter will affect the rate of expansion. So: How much was the expansion slowing? This question was what the two supernova teams had dutifully set out to answer, and they had succeeded: It wasn't.

The expansion wasn't slowing. The universe the two teams observed wasn't one where distant Type Ia supernovae were brighter than they should be at this particular redshift or that particular redshift, and therefore nearer. It was one where they were dimmer, and therefore farther. It wasn't a universe that was doing what an expanding universe full of matter acting under the influence of mutual gravitational attraction should be doing. It was doing the opposite.

The expansion of the universe was speeding up.

James Glanz broke the story in the 27 February issue of *Science*. Even though he had hinted at the possibility of a positive lambda in two previous articles, in October and January, such a result would be so difficult to accept that the community, rightly, was treating the possibility with what he considered "a preponderance of skepticism." An agreement between the two teams, however, might change that dynamic. Even before Filippenko's announcement at the UCLA meeting, Glanz had begun rounding up quotes from the High-z team. "To be honest," Bob Kirshner said, "I'm very excited about this result." Adam Riess said he was "stunned." Most quotable of all, from Brian Schmidt: "My own reaction is somewhere between amazement and horror." He elaborated: "Amazement, because I just did not expect this result, and horror in knowing that it will likely be disbelieved by a majority of astronomers—who, like myself, are extremely skeptical of the unexpected."

By the end of the day that *Science* published Glanz's article, 27

February, Riess had appeared on TV. (The national public television show's interviewer: "Why did some scientists react with what one called amazement and horror to these conclusions?") Articles in magazines and newspapers appeared around the world in the following week, culminating in a 1,600-word feature in the *New York Times* ("'My own reaction is somewhere between amazement and horror,' said Dr. Schmidt, the team leader").

At Berkeley Lab, the SCP team also responded with amazement and horror. Amazement because they had succeeded in finding nothing less than the fate of the universe, and horror because acceleration itself was what everyone suddenly wanted to talk about—and the High-z team was getting the credit for it. As Perlmutter said in a lab press release, the two teams were in "remarkably violent agreement."

"Basically, they confirmed our results," Gerson Goldhaber told the *New York Times*. "But they won the first point in the publicity game."

"Hey, what's the strongest force in the universe?" Kirshner said in the same article. "It's not gravity, it's jealousy."*

In early March, the High-z team submitted "Observational Evidence from Supernovae for an Accelerating Universe and a Cosmological Constant" to the *Astronomical Journal*.† The first week of May—a few days before the paper was even officially accepted—Fermilab convened a conference on the two supernova teams' results. An informal poll of the sixty or so attendees showed that forty were willing to accept the evidence.

Part of the rush to consensus in the community was sociological. Corroboration for any scientific result is always necessary, and if only one team had reached a surprising result, the response of the community would have been intense scepticism. That two teams had independently arrived at the same conclusion was notable. So, too, that the two teams had used mostly independent sets of data (very few of the same supernovae), had relied on several independent methods

---

* Actually, gravity is the weakest of the four forces. But "It's not the strong nuclear, it's jealousy" doesn't really land, as they say in standup comedy.

† A poke in the eye, or at least a joke: The other option was the *Astrophysical Journal*, but the High-z team wanted to underscore that they were doing *astronomy*.

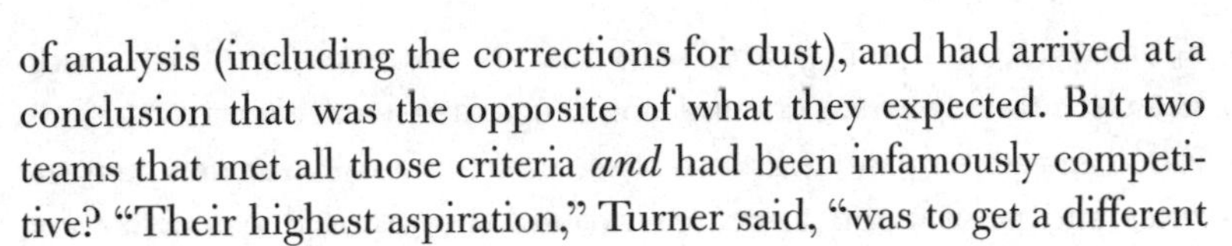

of analysis (including the corrections for dust), and had arrived at a conclusion that was the opposite of what they expected. But two teams that met all those criteria *and* had been infamously competitive? "Their highest aspiration," Turner said, "was to get a different answer from the other group."

And part of the rush to consensus was aesthetic. Just as inflation in 1980 solved the flatness and horizon problems, a positive lambda in 1998 made the universe understandable again. As Turner said, "It made everything fit!"

Those measurements of the Hubble constant on the "wrong" side of 60 that displeased Allan Sandage because they suggested a universe younger than its oldest stars? Problem solved. Those large-scale structures of supercluster filaments that seemed too mature for such a young universe? Problem solved. The universe was "too" young only if you assumed that the expansion rate had been decelerating throughout the history of the universe, or at least holding steady. A car that had been accelerating from 50 to 60 miles per hour and was only now reaching 65 would have needed more time to cover the same stretch of road than a car that had already been cruising at 65 miles per hour or slowing down from 70. If the expansion were decelerating, hitting the brakes, it would have been going *faster* in the "recent" past, and therefore taking *less time* to reach the present, than if it had just been constant. But an expansion that was accelerating today, hitting the gas, going faster and faster, would have been going *less* fast in the recent past, taking *more time* to reach the present. Thanks to acceleration, the age of the universe seemed to be, roughly, in the range of fifteen billion years, safely in the older-than-its-firstborn, old-enough-to-have-mature-filaments range.

But what made the supernova results perhaps most aesthetically pleasing wasn't just the presence of a positive value for lambda but the value itself.

If you were Bob Dicke or Jim Peebles in the late 1970s and you wanted the observation of a uniform cosmic microwave background to make sense, you wanted a theoretical explanation for homogeneity and isotropy. And then you got one: inflation. If you were Mike Turner or Rocky Kolb in the 1980s and you wanted the theory of inflation to work, you wanted an observation that revealed a flat universe. And then you got one—or half of one, anyway. COBE indicated that inflation was correct, meaning that the universe had to have an omega of 1. But numerous other observations indicated that the amount of mass in the universe was less than critical density, meaning that omega had to be less than 1—significantly so.

But now lambda explained away that contradiction. The amount of matter in the universe wasn't enough to halt the expansion, but the amount of matter *and energy* in the universe was. According to Einstein, matter and energy are equivalent, so while the mass, whether in the form of dark matter or regular matter, might well fall short of the critical density, the energy causing the acceleration—lambda—could make up the difference. A mass density of 40 percent or so plus an energy density of 60 percent or so added up to 100 percent of the critical density, or an omega of 1.

The universe *did* have a low matter density.

The universe *was* flat.

"Admit it," Jim Peebles once teased Brian Schmidt, "you didn't know what you had. You'd never heard of inflation."

"Inflation!" Schmidt answered. He informed Peebles that at Harvard he'd shared an office with the theorist Sean Carroll while Carroll was writing "The Cosmological Constant," one of the influential pre-1998 papers that explained how lambda could save inflation. "Alan Guth used to drop by once a week!" he added.

A positive lambda solved so many problems that when Turner approached Peebles in Aspen in 1998, he already knew what he wanted to debate: "Is Cosmology Solved?"

It was a debate Turner had already been framing. That March, for a dark-matter workshop in Gainesville, Florida, he had titled his talk "Cosmology Solved? Maybe." The following month, for a conference in Kyoto, Japan, he had dropped the qualifier from the title and went with a more straightforward "Cosmology Solved?" The published versions of both papers included the same sentence in the abstract: "These are exciting times in cosmology!" For the Smithsonian debate he took the exclamation point out of the body of the talk and promoted it to the title: "Cosmology Solved? Quite Possibly!"

Peebles would have to handle "quite possibly not," which was fine by him. It wasn't in his nature to argue passionately for a specific side of an unresolved issue, if only because having convictions about unresolved issues was unscientific. If he felt passionately about anything, it was that, in the absence of facts, you shouldn't feel too passionately. Fourteen years after he himself had used inflation's prediction of a flat universe as the basis for a lambda argument, Peebles still thought the community's embrace of inflation was premature—"distasteful," even. When he thought about physics, he divided its practitioners into classicists and romantics. The classicists were inventive but followed the rules; the romantics were respectful of the rules but followed their intuition. The romantics waved their hands and came up with a homogeneous, isotropic universe, and then, if they were lucky, an observation came along that could test their assumptions and predictions. A classicist looked to that observation—the one that suggested an expanding universe—then made a prediction of a temperature for the background radiation that observations would test. Then it was the romantics' turn again, waving their hands and invoking inflation and dark matter and, now, "missing energy"—the explanation that some classicist was going to have to invent for whatever physical presence in the universe corresponded to a positive lambda and caused the expansion to accelerate. Jim Peebles liked to think of himself as a classicist.

Mike Turner liked to think of Jim Peebles as "half enthusiast, half curmudgeon."

The debate took place on a wet Sunday afternoon in October at the National Museum of Natural History on the National Mall in Washington. The setting was Baird Auditorium, the same hall where the astronomers Heber Curtis and Harlow Shapley had "debated" in 1920 whether the Milky Way was the universe in its entirety or whether other "island universes" existed outside of it. Back then, Vesto Slipher's spectroscopy showing redshifts of nebulae was less than a decade old. Einstein's cosmology was only three years old and still applied to a static universe, thanks to his insertion of lambda. Hubble's discoveries that the nebulae were separate island universes and that, when their distances were graphed against their velocities, they seemed to be receding, lay in the decade to come—and with them, the apparent obsolescence of lambda. But now, some seventy years later, lambda was back. On their way into the auditorium, audience members received buttons bearing "Λ." If they were bewildered by the symbol, they weren't for long.

Like the earlier debate, the 1998 version wasn't going to solve anything; its purpose was to educate and entertain. And in terms of showmanship, as Peebles would have known in advance from having attended numerous talks by Turner, the debate was over before it began. All Turner had to do to win over the audience was to display one of his usual colorful viewgraphs, complete with Keith Haring–like dancing silhouettes:

> COSMOLOGY is EXCITING! . . . for at least the next 20 years
> STRONG FOUNDATION: Hot Big-bang Model
> BOLD IDEAS DEEPLY ROOTED IN FUND. PHYSICS: Inflation + CDM
> FLOOD OF DATA

(And all Turner had to do to make Peebles wince was say the words "precision cosmology.")

When the audience left four hours later, the drizzle might have felt like exclamation points dancing over their heads. But the question "Cosmology Solved?" was, by Turner's own admission, "ridiculous." As he acknowledged at the end of the debate, he was being "purposefully provocative." Debates might need a yes-or-no question, but Turner couldn't answer "Yes" and Peebles couldn't answer "No" without seeming foolish.

In a way, their roles on stage, while suiting their personalities, were almost perversely reversed. Turner argued, "I believe we will ultimately refer to 1998 as a turning point in cosmology as important as 1964"—the year that Wilson and Penzias inadvertently detected the cosmic microwave background at a temperature that Peebles himself had predicted. He cited the progress in establishing the most fundamental numbers in cosmology—the two that Sandage had always cited, plus the third that inflation had introduced. Astronomers were converging on a Hubble constant in the mid-60s. They were agreeing to a matter density of 0.4, give or take. And despite that seeming shortfall, they had discovered observational evidence that bumped the ratio between the *overall* density and the critical density—between the matter/energy density and the density necessary to keep the universe from collapsing—up to 1.

Yet Turner himself acknowledged that there was a problem that a positive value of lambda didn't solve. Cosmology had a new syllogism: One, the expanding universe was full of matter attracting other matter through gravity; two, the expansion was speeding up; therefore, something other than matter, dark or otherwise, had to be overwhelming the influence of gravity. So: What was it?

Cosmology solved? Hardly!

To astronomers, lambda was just a fudge factor, a symbol in an equation. It might equal zero. It might not. But if you had confidence in the usefulness of Type Ia supernovae for cosmology, and if you satisfied yourself that you'd checked your results, then you accepted its value. Brian Schmidt had been aware of the implications of a positive lambda for the theory of inflation, but Adam Riess, for instance, had not. In the days after Riess's computer code told him that the universe had negative mass unless he balanced it with a positive lambda, he'd had to educate himself—happily—about all the problems that a cosmological constant would settle.

For particle physicists, however, a positive lambda didn't solve a problem. It created one.

From a particle physics perspective, lambda wasn't just a number. It was a property of space. And space, in particle physics, wasn't empty. It was a quantum circus, a phantasmagoria of virtual particles popping into and out of existence. Not only did those particles exist, as experiments had shown, but they possessed energy. And energy, in general relativity, interacts with gravity. The result of quantum particles possessing energy that interacts with gravity was what physicists called the Casimir effect, after the Dutch physicist Hendrik Casimir. Put two parallel plate conductors closer and closer together, Casimir proposed in 1948, and you could measure the increase in the vacuum energy. Numerous experiments since then had found agreement with his predictions. As the mathematician Stephen Fulling noted, "No worker in the field of overlap of quantum theory and general relativity can fail to point this fact out in tones of awe and reverence."

So positive energy itself wasn't a surprise. And theorists even had two forms of vacuum energy in mind—or two names for them, anyway. One form of vacuum energy would be constant over space and time, and they would call it the cosmological constant. Another would vary over space and time, and they would call it quintessence (the fifth element in ancient Greek physics). In order to discourage astronomers from assuming that the terms "lambda" and "cosmological constant"—which they'd been using nearly interchangeably—were identical, Turner started testing other terms. "Funny energy" he auditioned at the Fermilab conference in May 1998, but that didn't stick. His next try—"dark energy," with its deliberate echo of "dark matter"—did.

The problem with the supernova result of a positive energy density in the universe, however, was that quantum mechanics predicted a value larger than the 0.6 or 0.7 that astronomers measured. A lot larger. Ten-to-the-power-of-120 larger. That's of 1,000,000,000,000,000,000,000,000,000,000,000,000,000,000,000,000,000,000,000,000,000,000,000,000,000,000,000,000,000,000,000,000,000,000,000,000,000,000,000,000 times larger. As the joke went, even for cosmology that's a big discrepancy. The stretching of space under the influence of such a ridiculously large energy density would be so extreme that, as Turner said, "you wouldn't be able to see the end of your nose." Not that the universe would be here for your nose to have an end on: A density that high would have cooled the cosmic microwave background below 3 K in the first 1/100,000,000,000,000,000,000,000,000,000,000,000,000,000,000th of a second after the Big Bang. So given the choice between an energy density with a value of $10^{120}$ and one with a value of 0.7, most particle physicists would have been perfectly content to assume that somehow someone someday would manipulate the maths, or figure out how particles were annihilating one another in just the right proportion, in order to make the result be what everyone had always been perfectly content to assume it was: $\Lambda = 0$.

Sceptics liked to quote a saying: "You get to invoke the tooth fairy only once"—meaning dark matter—"but now we have to invoke the tooth fairy twice"—meaning dark energy. An epithet became inescapable, or at least a commonplace on the conference circuit: epicycles. Were astronomers and their inflation-theorist enablers simply saving the appearances, like the ancients and their desperate measures to make the maths correspond to the motions in the heavens?

One could cite modern precedents. Scientists in the nineteenth century figured that the phenomenon of waves of light propagating across space like waves across water didn't make sense unless they inferred the presence of a cosmic pond. The Scottish physicist William Thomson, eventually Lord Kelvin, spent the entirety of his career trying to find equations to describe this "ether." In 1896, on the occasion of the golden jubilee of his service to the University of Glasgow, he wrote to a friend, "I have not had a moment's peace or happiness in respect to electromagnetic theory since Nov. 28, 1846." He died in 1907, two years after Einstein established the theory of special relativity by eliminating the need for absolute space, thereby making the ether "superfluous."

Would future generations look on the whole of modern cosmology as a similar lesson in the limitations of inferences from indirect evidence? The motions of galaxies didn't make sense unless we inferred the existence of dark matter. The luminosities of supernovae didn't make sense unless we inferred the existence of dark energy. Inference can be a powerful tool: An apple falls to the ground, and we infer gravity. But it can also be an incomplete tool: Gravity is . . . ?

Dark matter is . . . ?

Dark energy is . . . ?

Astronomers might not have been able to identify dark energy, but some theorists knew what it was: an inference too far. Just because a positive lambda would solve many problems didn't mean it existed.

"You observational astronomers," a theorist told Alex Filippenko in 1998, "are wasting a lot of valuable Keck and Hubble time, because your result must be wrong. We have no theory that could be compatible with a tiny non-zero vacuum energy"—tiny in the sense that lambda would be equal to 0.6 or 0.7 of critical density, rather than $10^{120}$—"and there's no theory that could possibly be compatible with this."

"Look," Filippenko said, "this is an observational result. I only know what end of the telescope to look through. You're a lot smarter than I am. But with additional observations, we will either confirm this, or we will find that we were wrong—hopefully for some subtle reason, and not '2 plus 2 equals 5' in some computer program."

In other words: Just because a positive lambda created a problem didn't mean it didn't exist.

In the end, sociology—the fact that two intense rivals had independently reached the same surprising result—wasn't going to be enough to convert the sceptics or, for that matter, to convince appropriately cautious astronomers that they weren't fooling themselves. Neither would aesthetics—whether the result solved problems or created problems. Not even the honor of being *Science's* "Breakthrough of the Year" for 1998. Filippenko's point was that only science, only further observations, could test a positive value for lambda.

And so astronomers did what scientists do in such circumstances: They set out to prove that the effect didn't exist. What problems might they have overlooked that could cause distant supernovae to appear dimmer than they should? Two possibilities immediately presented themselves.

One was an exotic kind of dust. Astronomers knew that regular dust within galaxies makes the light redder, and they knew how to correct for that dust—thanks in large part to Riess. His paper on the MLCS—multicolor light-curve shapes—correction method for dust won the 1999 Trumpler Award, an honour that recognizes a recent PhD thesis of unusual importance to astronomy. But now astronomers were mentioning the possibility of gray dust, and positing its presence between galaxies.

"Nobody has ever seen gray dust between galaxies," Riess thought. "But," he reminded himself, "nobody has ever seen a cosmological constant either."

Or what if the unusually faint appearance of supernovae at great distances was the result of supernovae being different back then, when the universe was younger and less complicated? What if the nature of Type Ia supernovae had changed over the life of the universe, and the recipe for a relatively nearby supernova was different from the recipe for a distant supernova? Maybe more distant supernovae had a simpler cocktail of elements, making them intrinsically fainter and giving the illusion that they were more distant.

There was one way to find out. If the interpretation of the supernova evidence was correct, then we were living at a time when dark energy was dominant over matter; the anti-gravitational force of dark energy was winning a tug of war with the gravitational force of matter. In that case, the expansion of the universe would be accelerating, and, as the two teams found, distant supernovae would appear dimmer than we would expect.

In earlier eras, however, the universe would have been smaller and therefore denser. The earlier the era, the smaller and denser the universe; the denser the universe, the greater the cumulative gravitational influence of matter. If astronomers could see far enough across the universe—far enough back in time—they would reach an era when dark matter was dominant. At that point, the gravitational influence of dark matter would have been winning the tug of war with the antigravitational force of dark energy. The expansion would have been *decelerating*, and supernovae from that era would therefore appear *brighter* than we would expect.

Not so the supernovae that we would see through gray dust, or that had a simpler cocktail of elements in the early universe. Those supernovae would just keep appearing fainter and fainter, the farther and farther we looked.

To distinguish between the two scenarios—dark energy versus either gray dust or changing cocktail—you would need to observe a supernova distant enough that it had exploded during that far earlier, far more distant era. You would need a supernova that had exploded before the expansion of the universe "turned over"—before the universe had made the transition from deceleration to acceleration, back when matter, not energy, was winning the tug of war. You would expect that supernova to be brighter than it "should" be. Plot it on the Hubble diagram—way out there, far beyond the nearby supernovae from Calán/Tololo, beyond the high-redshift supernovae that the two teams had discovered—and the slight upward deviation from the 45-degree straight line that High-z and the SCP had graphed would "turn over," too, just like the universe. It would dip down.

And if it didn't, you'd have to rethink dark energy.

Ground-based telescopes, however, couldn't see that far across the universe. The Hubble Space Telescope could, and it could even discover supernovae at that distance. From 23 to 27 December 1997, Ron Gilliland and Mark Phillips had used HST to try to prove that you could do just that—detect supernovae from the earliest epochs of the universe. For their search, they chose a familiar, even famous, speck of sky: the Hubble Deep Field.

Two years earlier, in 1995, HST had made the most distant image of the universe. For ten days the telescope had drilled a hole in the sky the size of a grain of sand at arm's length, just soaking up photons, seeing deeper and deeper across space and therefore farther and farther back in time. In the end the Hubble Deep Field contained about three thousand galaxies, some faint blue and among the first in the universe. Gilliland and Phillips wanted to make a repeat visit and do what supernova hunters had been doing since the 1930s—compare the earlier images with a current image and see what had changed. Did any of the galaxies in 1997 contain a speck of light—a supernova—that hadn't been there two years earlier?

Two did. Those specks got the designations SN 1997ff and SN 1997fg. Without follow-up observations, Gilliland and Phillips couldn't do the photometry that would allow them to construct light curves. But they'd made their point. You could use HST to discover supernovae at distances inaccessible from telescopes on Earth.

What you couldn't do, however, was what astronomers needed to do in order to test dark energy: make multiple reference images, then return to the same field in weeks to come in the hope of discovering the most distant supernova yet, for which you would have already reserved time for follow-up observations in the weeks and months ahead. You couldn't guarantee that you wouldn't be wasting HST time.

Still, those two supernovae—SN 1997ff and SN 1997fg—bothered Adam Riess. He couldn't stop thinking about them. By 2001, while remaining a member of the High-z collaboration, he was a staff scientist at the Space Telescope Science Institute, so HST results and possibilities were always on his mind. But SN 1997ff and SN 1997fg were a particularly poignant reminder of a lost opportunity. They were at a distance sufficient to test the deceleration-before-acceleration period of the dark-energy cosmological model—when the expansion was still slowing down under the dominant influence of dark matter, rather than speeding up under the dominant influence of dark energy. If only Gilliland and Phillips had been able to do follow-up work on SN 1997ff or SN 1997fg, Riess thought, astronomy would already have been able to put dark energy to a particularly compelling test.

In early 2001, Riess realized he could reframe the question.

What if one of those supernovae *had* been followed up? Not deliberately, by Gilliland and Phillips, but serendipitously, by HST during some other observation?

So he called up the HST search page on his office computer. He typed in the coordinates. Right ascension: 12h36m44s.11. Declination: +62o12'44".8. He requested dates that would correspond to the period during which SN 1997ff and SN 1997fg would have brightened and dimmed: December 27 1997 to April 1 1998.

Riess understood that the possibility of his finding what he wanted was extremely remote. HST looked at a lot of space; what were the chances that it had been staring at a particular dot of deep space during a particular period of time?

"Nobody's that lucky," he told himself.

Adam Riess was that lucky.

Earlier in 1997, Space Shuttle astronauts had added a couple of instruments to HST, including the Near Infrared Camera and Multi-Object Spectrometer, or NICMOS. By seeing in the infrared, NICMOS was particularly sensitive to distant objects whose light was so redshifted that, by the time it reached our patch of the universe, it had left the visible part of the electromagnetic spectrum. The NICMOS team had decided to test their instrument on a particularly distant patch of space with features they could easily identify: the Hubble Deep Field.

The observation program didn't begin until 19 January, but the camera took some test images in the interim. And there it was: 1997ff. It was in the HST archives of the NICMOS test run, on 26 December, 2 January, 6 January. Once NICMOS started taking data for real, 1997ff appeared in frame after frame, right at the edge. Sometimes it fell off the edge. But usually it was there. Riess spent the early part of 2001 examining SN 1977ff, establishing from the redshift that it had exploded about 10.2 billion years ago—far earlier than the period when the expansion of the universe would have gone from slowing down to speeding up. If indeed the universe *had* gone from deceleration to acceleration. If indeed dark energy existed.

Every spring the Space Telescope Science Institute hosted a symposium. One previous symposium topic had been the Hubble Deep Field; another had been stellar evolution. The topic in 2001 happened to be "The Dark Universe: Matter, Energy and Gravity." It was a chance for more than a hundred astronomers from around the world to reflect on their seemingly oxymoronic mission—what the symposium organizer and astrophysicist Mario Livio called "astronomy of the invisible."

STScI occupied a low, modern—and somewhat modest, considering its NASA provenance—building on a winding road in a far corner of the Johns Hopkins University campus in Baltimore.* It looked as if it were ducking its head so it could fit under the trees. At the rear of the building, outside the door to the auditorium on the ground floor was a wall of glass overlooking a brook. No Fermilab-style stampedes-around-the-accelerator-track-followed-by-a-barbecue-for-hundreds here; wine and cheese or the occasional overcaffeination was as wild as an STScI meeting would get.

By now, even the most ardent dark-energy sceptics had learned to accommodate the findings from a series of balloon experiments that had been launched from the outskirts of Antarctica and the Atacama Desert in Chile. The balloons had floated to an altitude of 30,000 metres and scraped the underbelly of outer space, at which point the onboard detectors had surveyed the cosmic microwave background. The goal was to refine COBE's measurements of the differences in temperature between points on the sky. If the differences in temperature were greatest between points separated by less than 1 degree, then the universe was open; by more than 1 degree, then the universe was closed; by 1 degree, then the universe was flat. So far, the verdict was all flat.

But saying that the universe sure looked flat wasn't quite the same as saying that the expansion of the universe was accelerating. You couldn't rely on an argument by subtraction—an omega of 1 minus a mass density of 0.3 equals a lambda of 0.7. The maths showed only the same seeming paradox that had existed pre-1998: an apparently flat universe via COBE, an apparently open universe via other observations. The balloon experiments made COBE's flat universe much more compelling, to the point that a flat universe was quickly becom-

* Checking into a local hotel on a previous visit, I asked the desk clerk where I could find the headquarters for the Hubble Space Telescope. He disappeared into a back room to confer with a supervisor, then returned to the counter. "The Hubble Space Telescope is"—he paused, and pointed—"up there."

ing cosmological orthodoxy. But acceleration? Especially if you were a particle physicist, that result still didn't make sense—still left you pining for alternatives.

Among the observers in attendance was Vera Rubin, opening the conference with a historical overview of dark matter—or, actually, a historical overview of the *idea* of dark matter, since, as she pointed out, until you know what dark matter is, you can't really know its history. She recalled predicting in 1980 the discovery of dark matter within ten years, and she said she was amused to see the British astronomer Martin Rees recently making the same prediction. She said she knew what Fritz Zwicky would have said about the current state of cosmology: "Epicycles!"

Among the theorists in attendance was Michael Turner, exhorting the congregation to indulge in "irrational exuberance" and embrace the era of "precision cosmology." To a fellow theorist complaining about the $10^{120}$ problem, Turner responded with exasperation: "Can't we be exuberant for a while?"

Saul Perlmutter was there too, talking up the possibilities of a space telescope dedicated to supernovae, and a couple of dozen other presenters were there to promote their own prospective research projects and report on their latest observations and postulate extravagant possibilities about the identity of dark energy. But mostly everybody was there to try to answer the question that Mario Livio had written on a transparency for his talk summarizing the symposium: "Accelerating Universe—Do We Believe It?"

Which is why it was Adam Riess who stole the show.

He choreographed his presentation, on the third day of the four-day conference, as a striptease. He had to do *something* to spice it up, since everybody in the auditorium knew what he would be showing. Two days earlier, on the first day of the symposium, Riess had attended a NASA press conference in Washington, D.C. to announce his discovery. And one day earlier, that announcement had made the front page of newspapers around the world. Still, now was his chance to let his fellow cosmologists examine this new evidence for themselves.

He would be using an overhead transparency. He kept most of it covered at first, while he explained what he would be showing. It was a Hubble diagram—redshift against brightness—of the supernovae from both the SCP and High-z teams. The points in this case represented not individual supernovae but averages of supernovae at similar redshifts.

Riess revealed the first three dots on the transparency: here, and here, and here, the averages of the nearby supernovae from the Calán/Tololo survey.

Then, moving to the right, the next three dots: here, and here, and here, the averages of the distant supernovae from the SCP and High-z searches.

The dots were beginning to describe the now-familiar gentle departure from the straight line, the upward turn toward the dimmer. In six dots Riess had taken his audience from a few hundred million light-years across the universe, to a billion, then two billion, three, four. Now, he said, he had the point that represented SN 1997ff. He had determined its redshift to be about 1.7, the farthest supernova to date by a long shot, a distance of about eleven billion light-years.

They knew what they were going to see, but the hundred or so astronomers in the auditorium couldn't help themselves. They shifted in their seats. Leaned forward. Held back. Crossed arms.

*There:* SN 1997ff.

A gasp.

The gentle upward curve was gone. In its place was a sharp downward pivot. The supernova was twice as bright as you would naïvely expect it to be at that distance. The universe had turned over, all right.

While Riess went on to explain that the result ruled out the hypothetical effects of exotic gray dust or a change in the nature of supernovae at a confidence level greater than 99.99 percent, the evidence continued to loom on the screen behind him. His audience couldn't take their eyes off it. For the astronomers of the invisible, it was something to see.

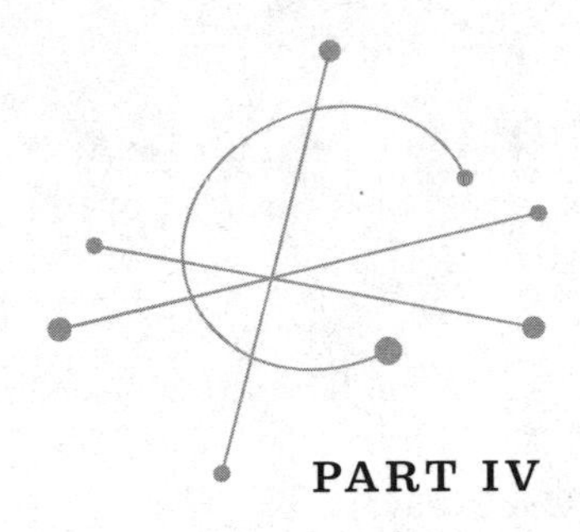

PART IV

# Less Than Meets the Eye

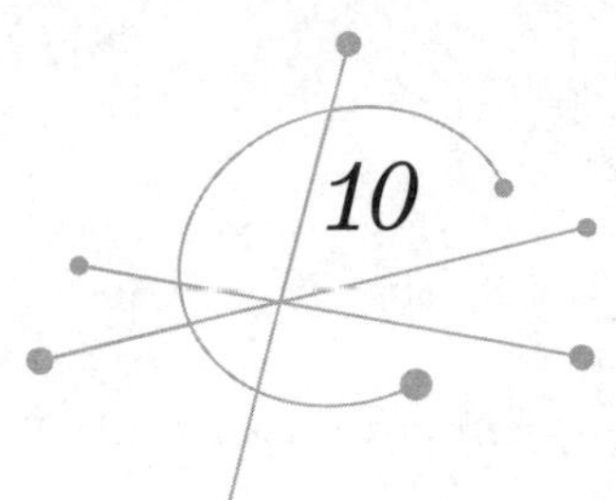


# The Curse of the Bambino

"I'M JUST GOING to watch this for a little while."

"Because of the funny noise?"

"Because it stepped backwards."

"It stepped backwards?"

"It stepped backwards."

"It stepped backwards." A pause. "That's impossible."

The two graduate students were staring at a pinkie-sized shaft—a device that was turning gears that were turning gears inside a copper cylinder that extended some four metres underground. The shaft was rotating, or "stepping," clockwise in tiny *tick, tick, ticks*. A counterclockwise step might have been impossible, but the first student had seen it with his own eyes. Now he needed to see it again.

He jammed his hands in his pockets. Then he pulled his hands out of his pockets and crossed his arms. Next he leaned one hand against a concrete pole. Then he grabbed a swivel chair and rested a knee on the cushion. He didn't take his eyes off the shaft. Another intern wandered past, asked what they were doing, and joined the staring contest.

The shaft was the first to blink. After ten minutes it stepped backwards again.

The three students marched over to the indoor shack where the other members of the team were huddling in the air-conditioning. Their presence pushed the shack to its capacity: eight. The first graduate student announced his finding to Les Rosenberg, one of the leaders of the project. Rosenberg, bushy-bearded and balding, smiled, but not really.

"That's impossible," he said.

"Oh, it's just the software," said yet another member of the team, not even glancing up from a desktop computer.

Still, Rosenberg had to see for himself. Soon four physicists, hands in pockets, were staring at the shaft.

*Tick. Tick. Tick. Tick. Tick. Tick. Tick.*

And so went the search for dark matter one summer afternoon in 2007 in a tin-roofed hangar in the California desert forty miles east of the Bay Area—officially, Building 436 of Lawrence Livermore National Laboratory, but more commonly, "the shed." The experiment was state-of-the-art, though at the moment it was more state-of-the-workbench. The interns were working from blueprints they'd spread on the concrete floor, and they were variously wielding wire cutters and wrenches, drill bits and hammers and a hacksaw. Drips, dents, flakes, scrapes, and spills decorated the tables and metal shelves. The "To Do" list on the whiteboard hanging next to the shed entrance was numbered 1 to 8, though the 8 was on its side: infinity. After lunch, the software guy fixed the software glitch: infinity minus one. The experiment was nearly twenty years in the making—this incarnation of the instrument would be the second—and it had another decade or so to go. But in the end, after the experiment had run its course, the world would know whether one of the two leading candidates for dark matter actually existed.

Even as Vera Rubin and her galaxy-motion-measuring colleagues in the 1970s were converging on the evidence for "missing mass" and prompting cosmologists to ask the inevitable question *What is it?*, parallel developments in particle physics were coincidentally coming up with a possible answer: *Not the stuff of us.* Not the stuff of atoms—the protons and neutrons, collectively called baryons, that have been forming and re-forming familiar matter from the first instant of the universe. Other stuff instead, also left over from the first instants of the universe, but not forming and re-forming—not interacting with itself or any other matter. Stuff that was weighing down the universe just by being there in abundance, but not doing much else. In the 1970s theorists were coming up with these hypothetical particles by the bushel in an effort to solve some problems with the standard model of particle physics. But when they looked at the properties such particles would have, they noticed that two in particular would exist in exactly the right proportion to make up the amount of matter in the universe that was "missing."

One was the axion, the particle that the physicists in "the shed" were hoping to detect. If it existed, then it did so by the trillions per cubic centimetre, and several hundred trillion would be threading their way through your body right now. Physicists are used to the idea of particles passing through seemingly solid objects; a neutrino could pass through a light-year of lead without coming into contact with another particle. But as with the search for the other leading dark-matter candidate, called the neutralino, the trick with the axion was to catch it.

Karl van Bibber started chasing the axion in 1989, when he was still on the prodigy side of forty. Three years later he recruited Rosenberg, a former student of his at Stanford whom he considered "an absolute genius" and "a world-class experimentalist," to join him in the Axion Dark Matter Experiment (ADMX). Van Bibber grew up in Connecticut, a fan of the Boston Red Sox, who infamously hadn't won the World Series in baseball since 1918. He spent his childhood hearing about how the Red Sox sale of Babe Ruth to the Yankees in the 1919–20 off-season had cursed the ball club. When the Red Sox went to the World Series in 2004, van Bibber's enthusiasm proved contagious; Rosenberg, who already felt some fondness for the team from his years at MIT, joined his colleague in rooting for the Red Sox. Van Bibber's screen saver on his desktop computer at Liver- was a floating compendium of newspaper headlines from the Red Sox World Series victory in 2004: "Ghost Busters!" "BELIEVE IT!" "SEE YOU IN 2090!" He and Rosenberg agreed that being a Red Sox fan was good training for being an axion hunter.

To have any hope of catching an axion they had to build a radio receiver that could track a signal with a "strength" in the vicinity of a trillionth of a trillionth—or 1/1,000,000,000,000,000,000,000,000th—of a watt. That's three orders of magnitude fainter than the final transmission from the *Pioneer 10* spacecraft in 2002, when it was seven billion miles from Earth and well on its way out of the solar system. But with *Pioneer 10,* scientists at least knew the signal's frequency; they knew where to turn the radio dial.

Karl van Bibber didn't have that luxury. But he did have an advantage over other dark-matter hunters: He'd know his prey if he saw it.

How do you see something that is dark, if by "dark" you mean, as astronomers beginning in the 1970s and 1980s did, "impossible to see"? How do you do something that is, by your own definition, impossible to do?

You don't. You rethink the question.

For thousands of years, astronomers had tried to apprehend the workings of the universe by looking at the lights in the sky. Then, starting with Galileo, they learned to look for more lights in the sky, those that they couldn't see with their eyes alone but that they could see through a telescope. By the middle of the twentieth century, they were expanding their understanding of "light," looking through telescopes that saw beyond the optical parts of the electromagnetic spectrum—radio waves, infrared radiation, x-rays, and so on. After the acceptance of evidence for dark matter, astronomers realized they would need to expand their understanding of "look." Now, if they wanted to apprehend the workings of the universe, they would have to learn to look in a broader sense of the word: to seek, somehow. To come into some manner of contact with. Otherwise, they could do only what ancient astronomers had been compelled to do, in the absence of instruments that extended one of the five senses: Save the appearances. Think. Theorize.

And theoretical was all that dark matter was. From the start, the evidence for it was indirect. We "knew" it was there because of how it affected stuff we could see. The obvious answer to what it was, was more of the same—more of the stuff we *would* be able to see, if only it weren't so distant, or so inherently dim, that it foiled our usual means of observation. Ockham's razor argued for a universe that consists of matter we already know—matter made from baryons—not matter we don't. Maybe, as Vera Rubin liked to joke, dark matter was "cold planets, dead stars, bricks, or baseball bats."

In 1986, Princeton's Bohdan Paczynski suggested that if these massive objects we couldn't see did exist in the halo of our own galaxy—where astronomers thought most of the Milky Way's dark matter resided—we could recognize their presence through a technique called gravitational lensing. In 1936, Einstein had suggested that a foreground star could serve as a lens of sorts on a background star. The gravitational mass of the foreground star would bend space, and with it the trajectory of the light from the background star, so that even though the background star was "behind" the foreground star from our line of sight, we would still be able to see it. "Of course," Einstein wrote in an article, "there is no hope of observing this phenomenon directly." To the editor of the journal he privately confided, regarding his paper, "It is of little value."

Einstein, however, was thinking small. He was still stuck in the universe in which he'd come of age. But the universe was no longer swimming only in the stars of our galaxy; it was swimming in galaxies. A few months after Einstein published his brief paper on the subject, Fritz Zwicky pointed out that rather than a foreground star, a foreground *galaxy* could serve as a gravitational lens. And because a galaxy had the mass of billions of stars, "the probability that nebulae which act as gravitational lenses will be found becomes practically a *certainty*."

In 1979, that prediction came true when astronomers found two images of the same quasar thanks to the gravitational intervention of a galaxy. The advent of CCD technology and supercomputers, Paczynski realized, might allow astronomers to make gravitational-lens detections on the small scale that Einstein had described, then dismissed. Paczynski reasoned that if, from our line of sight, a dark object in the halo of our galaxy—a Massive Compact Halo Object, or MACHO—passed in front of a star in a neighbouring galaxy, the gravitational effect of the dark foreground object would cause the light from the background object to appear to brighten. In 1993, two teams reported that after monitoring the brightness of millions of stars in the Large Magellanic Cloud, they had likely observed three such events—an impressive exercise in astronomy, but not a rate of discovery that suggested a Milky Way halo teeming with dark and massive objects made of baryons.

Then again, maybe the problem wasn't some unobservable matter but the observable effect—gravity. In 1981, Mordehei Milgrom, of the Weizmann Institute in Rehovot, Israel, arrived at Modified Newtonian Gravity, or MOND—a mathematical formula that he claimed described the light curves for galaxies just as well as, and probably better than, the presence of some sort of mystery matter. It did not, however, describe galaxy clusters very well.

But even if it had, physicists had already recognized a seemingly less obvious yet, somewhat paradoxically, more persuasive solution to the dark-matter problem than either the stuff we know or modified gravity: stuff we don't know.

As part of his inner space/outer space research, David Schramm as well as his students had discovered that deuterium (an isotope of hydrogen that has one neutron in the nucleus instead of none) could only be destroyed in stars rather than created (as other elements could be). Therefore, all the deuterium in the universe today must have been present in the earliest universe, and you could conclude that the present amount was at least the primordial amount. Through further calculations you could figure out how dense with baryons the early universe must have been in order for that minimum amount of deuterium to have survived that primordial period. The denser the baryonic matter, the steeper the drop in the deuterium survival rate. In order for at least this much deuterium to have existed in the early universe, the density of baryonic matter must have been at most a certain amount. This analysis therefore revealed an upper limit on the density of baryonic matter. (Schramm and Turner came to call deuterium a "baryometer.")

By similar reasoning and calculations, you could arrive at a lower limit for baryonic matter. Helium-3 (two protons plus a neutron) could only be created in stars rather than destroyed, so you could conclude that the present amount was at most the primordial amount. Then you could calculate how dense with baryons the early universe must have been in order for that maximum amount of helium-3 to have survived, and from that amount you could arrive at a *lower* limit on the density of baryonic matter.

By using particle physics to set upper and lower limits on the density of baryonic matter in the universe, Schramm and others converged on an omega for baryonic matter of about 0.1.

That amount, however, said nothing about non-baryonic matter.* Soon observations "weighing" the universe on different scales began

---

* And therefore also said nothing about a value of omega for the total amount of matter. So if you wanted a flat, omega-equals-1, inflationary universe, the number 0.1 for the ratio of baryonic matter to critical density wasn't particularly troubling. The ratio of total density — baryonic *and* non-baryonic matter — to the critical density could still be equal to or even greater than 1.

converging on a number of their own—an omega in the 0.2 range, and perhaps higher. That disparity alone—0.1 baryonic matter versus 0.2 total matter—provided evidence for the existence of more than black holes, bricks, and baseball bats in the halos of galaxies or suffusing galaxy clusters. The universe needed non-baryonic matter. And in a Big Bang model, such matter could come from only one place—the same place as the protons and neutrons and photons and everything else in the universe: the primordial plasma.

Even if particle physicists didn't know what these particles were, they knew that, like all the other particles that have been streaming through the universe since the first second of the universe, they had to be either fast or slow. Particles that were very light and moved at velocities approaching the speed of light—relativistic velocities—were called hot dark matter. Particles that were heavier and therefore more sluggish, attaching themselves to galaxies and moving at the same pace as the stars and gas, were called cold dark matter. And those two interpretations came with a crucial test.

In the early 1980s, astronomers hadn't yet detected the primordial ripples in the background radiation that would have corresponded to the so-called seeds of creation—the gravitational gathering grounds that would become the structures we see in the current universe. Even so, theorists knew that if those ripples did exist, then the two models of dark matter—hot and cold—would have affected them in different ways, leading to two opposite evolutionary scenarios for the universe.

Hot dark matter—particles moving at relativistic velocities—would have smeared the primordial ripples to large volumes, like a downpour on sidewalk chalk. In a universe full of matter gathering around those vast swaths, larger structures would have formed first. These vast gobs of matter would then have had to break up over time into the specks we see today—galaxies. The universe would have had a top-down, complex-to-simple history.

Cold dark matter—particles moving at a small fraction of the speed of light—would have sprinkled the primordial ripples much more subtly and affected the evolution much more slowly. Structure in that universe would have started as specks, or galaxies, and worked its way up to larger and larger structures. The universe would have had a bottom-up, simple-to-complex history.

The observations in the early 1980s indicating that the Milky Way is part of a Local Supercluster, or that superclusters are separated by great voids, provided enough support for the cold-dark-matter model that most theorists abandoned the hot-dark-matter model by the middle of the decade. Then astronomers began using redshift surveys to map the universe in three dimensions, beginning in the late 1980s with the dramatic Harvard-Smithsonian Center for Astrophysics sighting of a "Great Wall" of galaxies. From 1997 to 2002, the Two-degree-Field Galaxy Redshift Survey, using the 3.9-metre Anglo-Australian Telescope, mapped 221,000 galaxies; beginning in 2000, the Sloan Digital Sky Survey, operating on the 2.5-metre telescope at the Apache Point Observatory in New Mexico, mapped 900,000 galaxies.

In those surveys and others, astronomers found that the farther across the universe they looked—and therefore the farther back in time—the *less* complexity they saw. Which is another way of saying that the closer they got to the present, the *more* complexity they saw. Galaxies formed first, at redshifts of 2 to 4—or roughly nine to twelve billion years ago. Then those galaxies gathered into clusters, at redshifts of less than 1—or less than roughly six billion years ago. And now, today (in a cosmic sense), those clusters are gathering into superclusters. Matter clumped first in small structures, and those small structures continued to gather together. The universe has apparently had a bottom-up, simple-to-complex history, consistent with theoretical cold-dark-matter models.

Still, what those surveys mapped were sources of light. They showed where the galaxies were, leaving scientists to infer where the dark matter was. In 2006, the Cosmic Evolution Survey, or COSMOS, released a map of the dark matter itself. The survey studied 575 Hubble Space Telescope images of instances in which two galaxies or clusters of galaxies lined up one behind the other. Like the microlensing technique that the MACHO surveys had used, weak gravitational lensing relied on a foreground concentration of mass to distort the light from a more distant source. Unlike microlensing, however, weak gravitational lensing recorded not individual events, as objects passed in front of other objects, but ongoing relationships between objects that were, for all practical purposes, stationary relative to each other—galaxies or clusters of galaxies. The light from a foreground object told astronomers how much mass *appeared* to be there. The gravitational-lensing effect on the background object told them how much foreground mass *was* there. The difference between the two amounts was the dark matter.

The COSMOS map not only covered an area of the sky nine times the diameter of the full moon, but was three-dimensional; it showed depth. It was like the difference between a map that shows only roads and a map that also shows the hills and valleys that the roads traverse. And because looking deeper into space means looking back in time, the COSMOS map showed how those hills and valleys got there —how the dark matter evolved. According to this "cosmopaleontology," as the team called this approach, the dark matter collapsed upon itself first, and then those centres of collapse grew into galaxies and clusters of galaxies—again, an image consistent with the bottom-up, cold-dark-matter formulation.

Perhaps the most dramatic, and certainly the most famous, indirect evidence for the existence of dark matter was a 2006 photograph of a collision of two galaxy clusters, collectively known as the Bullet Cluster. By observing the collision in x-rays and through gravitational lensing, Douglas Clowe, then at the University of Arizona, separated visible gas from invisible mass. The visible (in x-ray) gas from both clusters pooled in the centre of the collision, where the atoms had behaved the way atoms behave—attracting one another and gathering gravitationally. Meanwhile, the invisible mass (detectable through gravitational lensing) appeared to be emerging on either side of the collision. It was as if dark-matter freight carriages from both clusters had raced, ghostlike, right through the cosmic train wreck.

The photograph appeared around the world, and the Bullet Cluster became synonymous with dark matter. The false colour helped: NASA assigned the visible gas pinkish red and the invisible mass blue. The headline on the press release also helped: "NASA Finds Direct Proof of Dark Matter."

But that wasn't quite true. Even leaving aside the dubious use of the word "proof," the "direct" was subject to debate—and had been closely parsed during the writing of the press release. The problem was that astronomers had been saying for a generation that dark matter dominated baryonic matter in the universe. Now they were saying that dark matter dominated baryonic matter in the universe. "It's not 'direct,'" Clowe conceded. "A true dark-matter direct detection would be catching a particle."

So how could you catch one? How could you capture the evidence that, as Mike Turner liked to say, "you could put in a bottle and bring to the aunt from Missouri who's saying, 'Show me'"? First, you would have to know what to look for — or "look" for.

By the late 1970s, theorists had finished fashioning the standard model of particle physics, an explanation of the relationships among three of the four fundamental forces in the universe — electromagnetism, weak interaction (or weak nuclear force), strong interaction (or strong nuclear force). The particles themselves came in two types, bosons and fermions — those that, respectively, can and cannot occupy the same quantum space. Some theorists proposed a "supersymmetry" between bosons and fermions; each boson would have a fermion partner, and vice versa. The photon, for instance, got a photino superpartner, the guage boson a guagino, the gluon a gluino. And the neutrino got a neutralino.

The neutralino — even before the axion or MACHO — turned out to be an attractive candidate for dark matter. Theorists' calculations predicted how many of these neutralinos would have survived to the present universe, and they predicted the mass of the neutralino, and when they added up those two numbers, the answer was nearly identical to the best estimates of the amount of dark matter. Aesthetically, physicists liked that the neutralino wasn't ad hoc; nobody invented it to solve the problem of dark matter. The neutralino would just be there, and its connection to dark matter was a bonus.

The trouble with the neutralino, from an observer's perspective, was that it interacted only through the weak force. Hence the name that Mike Turner bestowed on this class of dark-matter candidates: Weakly Interactive Massive Particle, or WIMP.* A WIMP wouldn't interact through electromagnetism, meaning that we couldn't see it in any wavelength. It also wouldn't interact through the strong nuclear force, meaning that it would rarely interact with atomic nuclei. The key word, though, is "rarely."

---

* The acronym preceded, and inspired, MACHO.

The very occasional exception was the opening that dark-matter detectives needed. It allowed them to take evidence that would be inaccessible to our senses and transform it into evidence that *would* be accessible. They still wouldn't be able to see the WIMPs themselves, but they would theoretically be able to see two after-effects of a WIMP–nucleus interaction. One would be a minuscule amount of heat from the agitated nucleus. The other would be an electric charge from loosened electrons. Neither of those after-effects in itself would be enough to identify a neutralino. But the combination of the two in a single event would be a signature unique to the particle.

To "look" for these effects, however, scientists would have to adopt another kind of "telescope," one that was new to astronomy: the laboratory.

One of the start-up programmes at the Center for Particle Astrophysics in the late 1980s (along with the experiment that would become the Supernova Cosmology Project) was an effort at this kind of detection, the Cryogenic Dark Matter Search, or CDMS. In order to stabilize the target atoms — germanium, in this case — the detector had to maintain a temperature of .04 of a degree Celsius above absolute zero. And in order to block out cosmic rays and other offending ordinary particles, the detector had to be shielded.

Under the leadership of the Center for Particle Astrophysics director Bernard Sadoulet, the CDMS project began life in a shallow site on the Stanford campus, 20 metres below ground, or roughly the equivalent of several hard turns in a subterranean car park. The problem wasn't getting a ping — a reading that showed an interaction with the nucleus of the germanium atom. Pings it got. The depth was sufficient to block out cosmic rays but not muons, which are like a heavy version of the electron. Muons penetrated the twenty metres of rock, hit the detector, and made neutrons, which leave a signal similar to the neutralino's but aren't, alas, neutralinos. The problem was getting the right *kind* of ping.

There was nowhere to go but down. In 2003, the successor detector, CDMS II, began operating under three-quarters of a kilometre of rock in a former iron mine in northern Minnesota. By then CDMS had inspired a generation of similar detectors, though the high cost, large scale, and long data gestation for CDMS prompted researchers to consider cheaper, faster approaches. Many of the second-generation detectors relied on the noble gases argon, neon, and xenon, which don't need to be cooled to anywhere near absolute zero to turn into a usable liquid form, and which are far less expensive. In 2007 the XENON10 experiment, a 15-kilogram tank of liquid xenon operating in the underground laboratory at Gran Sasso, Italy, established itself as a viable rival with the release of results at a far more sensitive level than CDMS II had yet been able to reach.

Back in 1992, Sadoulet had told a journalist, "I may be bragging, but I think we're close." Sixteen years, numerous rotations of graduate students and postdocs, and two generations of detector later, a group of twelve CDMS team members gathered at his home to await a "blind" analysis of their data—a test of whether the latest research they had done would coalesce into a quantifiable result. According to their calculations, over the preceding year the CDMS II particle detector should have registered no more than one or two "hits" from stray subatomic particles of ordinary matter. The fewer hits they saw, the more confidently they could eliminate a segment of WIMP phase space—the graph that showed all reasonable combinations of size and mass. Like the settlement of a frontier by pioneers, the elimination of each swath of the graph left a narrower region to explore. At precisely midnight, they gathered around a computer in Sadoulet's living room, "unlocked" the data, and waited for the answer to bloom into view.

Zero.

A cheer went up—not unlike the spontaneous applause that greeted a team member later that month at a UCLA symposium on dark matter when he stood before a hundred or so colleagues from around the world and re-created, via PowerPoint, the revelation of non-detection. CDMS II had leap-frogged back into the lead, leaving a XENON10 team member to interrupt his own PowerPoint presentation later that morning to sigh, "I guess this graph is about forty-five minutes out of date."

It was some indication of just how difficult the WIMP problem was that even a null result was cause for celebration. Later that day, one of the team leaders graciously accepted congratulations on his team's work as he boarded a lift. "Of course," he added, softly, as the doors closed, "a detection would have been better."

Nineteen months later, he got his wish. The next "unblinding party" for CDMS II was also its last. In the interim, that incarnation of the experiment—five towers of six detectors each—had been decommissioned to make way for an upgrade: SuperCDMS. When the team "opened the box" on that last round of data, they expected the result to be more of the same: plenty of nothing. Instead, they got two "somethings": one from 5 August 2007, the other from 27 October 2007.

A null result would have made a definitive statement, excluding one more phase space for future experiments to investigate. Two detections, however, occupied a particle physics purgatory. Statistically, that number wasn't enough even to claim "evidence for," let alone the "discovery" that five events would have justified. If both events were due to background noise such as cosmic rays or radiation from within the mine, then you were unlucky. If both events were indeed the "edge of the signal," and a competing collaboration such as XENON100 (the successor to XENON10, already up and running when CDMS II opened the box) wound up seeing a statistically satisfying number of events and got to claim the discovery . . . then you were still unlucky. As one graduate student said, expressing his disappointment at not getting a null result, "We would have totally dominated!"

"We're actually in the game to see something," Jodi Cooley had to remind him. The coordinator of data analysis for the experiment, she had joined the collaboration as a postdoc at Stanford five and a half years earlier, and she had secured her first faculty position, as an assistant professor at Southern Methodist University, two months earlier. By the standards of a Bernard Sadoulet, she was a newcomer to the dark-matter game. But she was also enough of a veteran that she had tired of celebrating the sighting of nothing.

Still, she knew what that grad student meant. In a way, Cooley told herself (though not the grad student), a total of two detections was "the worst-case scenario."

The collaboration spent the next few weeks running the results through data quality checks. Were the detections well inside the detector, where stray radiation was less likely to reach? Yes. Did a detection come during a time when the instruments had been behaving smoothly? Yes. Did a detection come at the same time as another detection — a double-WIMP detection that would have defied belief? No. Did the two detections occur on the same detector? No. In the end, the team subjected the results to more than fifty checks, and both detections passed every test.

The quality of the results was strong. It was the quantity that was the problem. The collaboration just didn't have enough events to let them know what they'd seen.

But they'd seen *something*. That fact alone made the result more worth reporting to the community than a null result would have been. The collaboration would have published a paper on the results no matter what the outcome, but this something — these two somethings — merited a more direct interaction with the community. The collaboration scheduled simultaneous presentations at Fermilab and Stanford the following month, as well as smaller educational sessions at other institutions involved in the collaboration. The subject of dark matter was tantalizing enough that the talks were sure to attract some attention. They had no idea.

Within days, rumours about the result were dominating the particle physics slice of the blogosphere. "Dark matter discovered?" "Has Dark Matter Finally Been Detected on Earth?" "Rumor has it that the first dark matter particle has been found!" "*¿Se ha descubierto la materia oscura en el CDMS?*" "*Pátrání po supersymetrické skryté hmotě.*" "みんな大好き (か、どうかは分かりませんが) dark matter を検出したという報告が出ています."

The team realized that by scheduling all the sessions for one day they had inadvertently given the impression that there was about to be a before-and-after moment in science. There wasn't. At best there would be a sort-of-before-and-sort-of-after-but-we-won't-actually-know-until-some-other-experiment-reinforces-our-results-and-even-then-today's-announcement-would-be-seen-in-retrospect-as-at-best-a-hint-of-detections-to-come moment. They "pre-poned" the Fermilab and Stanford announcements, moving them up a day, separating them from the more casual sessions, hoping to lower expectations.

Too late. If they did announce a null result, wrote one blogger, "the Thursday speakers will be torn to pieces by an angry mob, and their bones will be thrown to undergrads." To which "Anonymous" added in the Comments section on the same website, "Independent of the rumors,* I have it from a very well-known physicist that CDMS will in fact announce that they have discovered dark matter tomorrow." *Discover* magazine live-blogged Cooley's standing-room-only presentation, prefacing the tick-tock with: "Personally, I have heard rumors that they have either 0, 1, 3, or 4 signal events."

In fact, no, no, no, and no.

"THE NUMBER IS TWO!!!!"

Or as Cooley carefully explained, "The results of this experiment cannot be interpreted as significant evidence for WIMP interactions, but we cannot reject either event as a signal." Theirs wasn't a detection. It wasn't a null result. It was a neither–nor conclusion that taught the portion of the world that cared about such things a lesson that Jodi Cooley, Bernard Sadoulet, and a certain graduate student had already learned the hard way:

If you let it, dark matter will break your heart.

"I'm in love with the axion!"

Les Rosenberg didn't care who knew it. When the mood struck, he wasn't afraid to declare his affection to the world. Karl van Bibber—fit; never took the lift when stairs were an option; looked like Leonard Nimoy, in a good way—was a bit more discreet; the word he repeatedly used about his relationship to the axion was "smitten." Like the Red Sox—and unlike the WIMP—the axion seemed to inspire a certain kind of blind devotion and underdog identification.

The natural mathematical match between the neutralino and dark matter—how many neutralinos would have survived the primordial conditions, multiplied by the predicted mass of a neutralino, equaling the best estimates of the current density of dark matter—had always made it the favoured candidate among physicists. The longer it remained undetected, however, the more the community was willing to consider alternatives. The axion might not have been as obvious a match, but it was a match nonetheless.

---

* !

Like WIMPs, the axion was a hypothetical particle that fell out of an adjustment to the standard model. In 1964 physicists discovered the violation of a certain kind of symmetry in nature — in part, that the laws of physics wouldn't hold if a particle and its antiparticle traded places. In 1975 the physicists Frank Wilczek and Steven Weinberg independently realized that a particle with certain properties could solve the problem. "I called this particle the *axion,* after the laundry detergent," Wilczek once explained, "because that was a nice catchy name that sounded like a particle and because this particular particle solved a problem involving *axial* currents."*

Unlike WIMPs, however, the axion was not a massive particle. The neutralino would be fifty to five hundred times the mass of a proton; the axion would be one-trillionth the mass of an electron, which itself was 1/1,836th the mass of a proton. If axions existed, they would be a trillion times lighter than an electron, making the chance of their interacting — or coupling — with baryonic matter, as van Bibber said, "vanishingly small." But in 1983 the physicist Pierre Sikivie realized that while the axion, unlike the neutralino, couldn't couple with matter, it *could* interact with magnetism. Under the influence of a strong enough magnetic field, an axion could disintegrate into a photon — and *that's* what a detector could detect.

In 1989 van Bibber attended a meeting at Brookhaven National Laboratory, on Long Island, where Adrian Melissinos, of the University of Rochester, asked a couple of dozen physicists whether they wanted to participate in the construction of such a detector. He went around the table: "Are you in or out?" Van Bibber was in. When Melissinos had finished surveying the scientists, van Bibber pointed out that Melissinos hadn't polled himself. Was he in or out?

"This is too much like hard work," Melissinos said. "This is for you young guys."

And so van Bibber found himself leading an experiment that might outlive his professional life. ADMX was a highly magnetized resonant cavity. If axions were entering it, then they would interact with the magnetism and disintegrate into photons. If they disintegrated into photons, then they wouldn't be able to pass back through the casing

---

* Currents that move around the vertical axis.

of the cavity. Instead, they would remain inside, bouncing off the walls, emitting a faint microwave signal. That signal was what ADMX should be able to detect. In other words, ADMX was a radio receiver.

By 1997 he and Rosenberg had a prototype up and running. The following year they published a paper that put the community on notice: You could actually do Sikivie's strong-magnet, resonant-cavity axion experiment. Their impression was that the success of the prototype in 1998 surprised the community; it stunned them, anyway. That instrument—a waist-high copper cylinder—still sat in a corner of the shed. Once, the public relations department at Livermore contacted van Bibber about displaying it in the visitors' centre, and they asked him to send a photo. He did, and he never heard back.

But it remained a thing of beauty to him. As a kid, van Bibber wanted to be a cartoonist, just like his father, Max, who drew a popular comic strip called *Winnie Winkle*. Karl's drawings, however, were disturbing and his parents thought he might be emotionally troubled, until they realised he was colourblind. His father did nonetheless influence his choice of profession. One day he brought home from Manhattan a science textbook, and Karl, then in his early teens, performed experiment after experiment until he'd exhausted the book. He was "smitten."

For van Bibber, ADMX was a low-energy physics version of one of those experiments. It was usually a ten- to fifteen-person collaboration among friends, including the half-dozen kids who were doing the heavy lifting, literally. In collider physics, thousands of students could spend their entire educational careers writing the software for an experiment they might never be able to touch. "Cannon fodder," van Bibber called them. But here in the shed, the kids could spend their summers in T-shirts and shorts, griping about the heat. (Once, just for fun, the group moved a thermometer up a ladder one step at a time, and at each rung the temperature rose one degree, until it peaked at 48 °C.) And as they worked, they could find out whether they, like van Bibber, nonetheless got a kick out of building a detector that could find a signal equivalent to the cosmic microwave background . . . plus one photon.

ADMX was, in a real sense, a labour of love. Van Bibber loved that the axion was a high-risk career move. He loved that Rosenberg had the kind of "crazy streak" that allowed him to take that same risk. (For his part, Rosenberg called van Bibber the Mick Jagger of axions: the leader of the band, the salesman for the brand. And in 2006 the two of them even made the cover of *Rolling Stone*—or, at least, wrote a major article on ADMX for *Physics Today*.) Van Bibber loved that his collaboration was basically the only one in the world looking for the axion. He loved that the annual cost of the experiment was maybe 1 or 2 percent of the nearly sixty million pounds spent on the two or three dozen WIMP experiments underway around the world at any one time.

But most of all, van Bibber loved that the axion signal would be so unfathomably faint. It meant that he was performing a seemingly paradoxical feat: "macroscopic quantum mechanics." He loved that if the axion was there, the instrument would detect it. You wouldn't know the frequency in advance, so the search of the microwave spectrum would have to be numbingly methodical. But when Phase II was over—Phase I ended in 2004—he would *know:* The axion exists or it doesn't exist.

That certainty, van Bibber recognized, was something WIMP hunters could only envy—one of them a friend of his at the University of Chicago. Juan Collar was part of a generation that had joined the search for neutralinos in the 1990s. Since then he had abandoned the CDMS prototype. The acronym for his experiment, the Chicago Observatory for Underground Particle Physics (COUPP), which resided at a depth of 300 metres in a tunnel at Fermilab, was significant: The *p*'s were silent, as in "coup." Shaking his fist at an imaginary enemy, Collar would say, "It has the connotation of a terrible blow to the system"—the system being the whole cryogenic approach.

COUPP was less a technological advance than a throwback to an earlier era of physics: a bubble chamber. The chamber was filled with a superheated heavy liquid and outfitted with a camera; unlike other dark-matter experimenters, the COUPP team would have the thrill of seeing an actual visual result: a bubble. And bubbles they got: muons, again. Collar worried whether his generation—the particle physicists who had started out with such optimism in the 1990s, sure that they would be the ones to find the WIMP and win the race to discover dark matter—would stick around long enough to see the right kind of bubble, to hear the right kind of ping. He had his doubts. Sometimes at conferences Collar and his no-longer-on-the-prodigy-side-of-forty colleagues would convene at the hotel bar and "howl at the moon." And sometimes he would retreat to a downstairs lab at Chicago to play with a detector that had nothing to do with WIMPs, if only because when he put it next to a reactor, he would actually see a signal.

When Collar talked about his generation of researchers, he would say, "It gets kind of old, to look for a particle that might be there or not and always getting a negative result." A negative result from an experiment, after all, didn't mean that the neutralino didn't exist. It might mean only that theorists hadn't thought hard enough or that observers hadn't looked deep enough. Collar kept a graph taped to a wall in his office that showed the range where he and other researchers hoped the neutralino might reside, and sometimes he would find himself looking below the sheet of paper, at the blank wall. "If the neutralino is way down there," he would think, "we should retreat and worship Mother Nature. These particles maybe exist, but we will not see them, our sons will not see them, and their sons won't see them." And then he would think of his friend in the California desert. "Karl," Collar would tell himself, "knows he's going to get the job done, dammit."

But van Bibber, a generation older than Collar, had experienced a different kind of frustration: For years he and his fellow dark-matter hunters thought they owned the universe, if only they could find it. After 1998, they realized they owned maybe a quarter of the universe. Not bad, but van Bibber thought it was "sort of a rude demotion."

Yet, he remained sanguine, as someone still in love, long into a marriage, does. When he reached his mid-fifties and thought about the possibility that ADMX would take another ten years, and that he might wind up with nothing, and that it might be the experiment that would close out his career—that he would have spent the latter half of his professional life in one way or another looking for the axion—he thought it still would have been a worthwhile pursuit. He hoped, of course, that his experiment would be the one that detected dark matter. But sometimes when he and his old friend and longtime colleague Les Rosenberg got to talking, they had to admit that after the Boston Red Sox won the World Series, baseball was never the same.

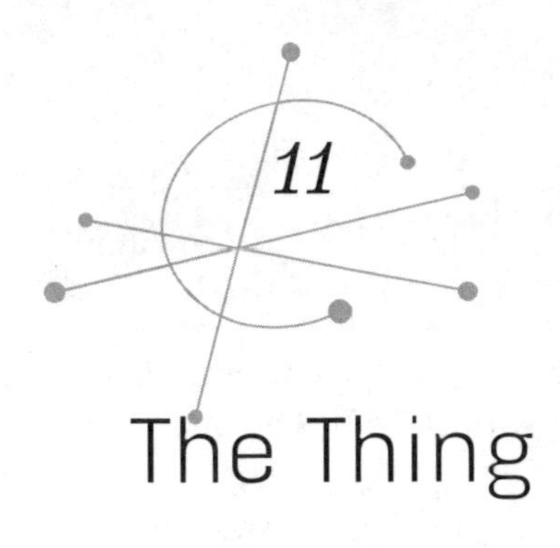

# 11 The Thing

THEY KNEW WHERE they were going. Or at least they knew where they hoped they were going, and they were pretty sure they were headed in roughly the right direction. Once in a while the wind would ease and the veil of snow would part and they would glimpse, in the distance, the distinctive silhouette of the Dark Sector. But then the wind would gather again, and the white would envelop them, and the summer crew for the South Pole Telescope would lower their heads and withdraw behind the fur lining of their hoods, trusting that they would soon be climbing the metal stairs to the laboratory and resuming their search for clues about dark energy, a mission that had now taken science to the ends of the Earth, literally.

Welcome to the most benign environment on the planet. Or so William L. Holzapfel, a UC Berkeley astrophysicist and veteran of several stays at the South Pole, liked to say, and not just because the whiteout was the exception and the recent weather had been unseasonably mild. Other days that week — the week between Christmas and New Year's, early summer in the Southern Hemisphere and midway through the six months that the Sun is continually up at the Pole — the temperatures were barely breaking past -20°C (and one day even broke above -17 to set a record high for the date), and the wind was mostly calm. Holzapfel routinely made the walk from the Amundsen-Scott Station (a snowball's throw from the Pole itself, which is marked with, in fact, a metal pole) to the telescope wearing jeans and trainers. One afternoon the lab's heating system system went a little haywire and the crew had to prop a door open to cool off. And for those of the two hundred fifty or so working at the

Pole who remembered to pack their swim trunks — including Holzapfel — the traditional makeshift New Year's Eve outdoor sauna was as bracing as ever.

But, from an astronomer's perspective, not until the Sun goes down and stays down — March through September, the austral autumn and winter, when temperatures drop to –70°C — does the South Pole get "benign."

For six months, telescopes at the Pole do nothing but swallow sky and funnel it north, and they do so while operating under impeccable conditions for astronomy. The atmosphere is thin; the Pole is more than 2,800 metres above sea level (of which the first 2,700 down are ice). The atmosphere is also stable, thanks to the absence of the heating and cooling effects of a rising and setting sun. And the area doesn't suffer whiteouts during the dark months; the Pole has the lowest peak wind velocity — 55 miles per hour — of any weather station on Earth.* But most important for the kind of astronomy performed by the South Pole Telescope, the air is exceptionally dry.

Technically, the Pole has a desert climate. Snowfall is rare. (The snow that's there is the result of millions of years of windblown accumulation from the periphery of the continent.) Chapped hands at the Pole can take weeks to heal, and perspiration isn't really an issue. The level of moisture is so low that if you took all the water vapour in the atmosphere at any one time during the coldest months and compressed it, the sheet would be less than three-tenths of a millimetre thick. For astronomers working at microwave or submillimetre wavelengths — cosmic microwave background researchers such as Holzapfel — the less water vapour the better. Even small amounts of atmospheric moisture can absorb submillimetre-wavelength signals, meaning that those CMB photons wouldn't even reach the telescope. That same water vapour can also emit its own submillimetre-wavelength signals, meaning that observers could find themselves mistaking humidity for history.

For Holzapfel, the SPT was only the latest in a series of CMB detectors at the Pole. When Holzapfel was a young PhD student at

---

* In contrast, Mount Washington, New Hampshire, at an elevation 900 metres lower than the South Pole, long held the record for highest wind measured at the surface of the Earth: 231 mph.

Berkeley, in the late 1980s and early 1990s, he would see the standard computer simulations of the CMB and think, "That's a nice story." The simulations would show the temperature that had to be there, if the Big Bang theory was correct. They would show the fluctuations in the radiation that had to be there, if inflation was correct. To Holzapfel, these simulations were targets — ideals that future data, when instruments were sensitive enough, could hope to approximate.

Then came COBE.

"Astonishing," thought Holzapfel. Forget about approximations. The match to the simulations was exact, at least within the (very) narrow margins of error. Not only did the CMB have a story, it was a story that Holzapfel himself wound up helping to tell. In a series of CMB surveys at the South Pole, the agreement between ideal simulations and actual data had become so fine that scientists couldn't adjust a variable without destroying the universe. Tweak the density of the dark matter in a CMB simulation even slightly, and it no longer matched the data. Leave the dark matter alone and tweak the dark energy instead, and again the simulation departed from the data. Do the same with the density of baryons, or the expansion rate of the universe; same thing happened.

In a way, the South Pole Telescope was one more CMB study. Like the dozens of other dark-energy experiments that had arisen in the first years of the twenty-first century, the SPT would be studying how the universe had changed over time — the evolution of its structure from those seeds in the CMB to what we observe today. This time, however, the story Holzapfel was hoping to tell wasn't only what the universe was, way back when, and how it got to be the way it is today. Instead, the SPT would be "seeing" into the future. In effect, the part of the story of the universe that Holzapfel would be helping to write was the end.

The fate of the universe? Again? Hadn't the Supernova Cosmology Project and the High-z team settled that question in 1998?

Not exactly. The universe, as it happens, was not as simple as they thought.

Even while the SCP team's acceleration paper, "Measurements of Ω and Λ from 42 High-Redshift Supernovae" — submitted to the *Astro-*

*physical Journal* in September 1998 and published the following June—was making its way through internal revisions and peer review, Saul Perlmutter was thinking about the next step in the supernova game. How could you get the greatest number of supernovae at the highest redshifts? The obvious answer: a space telescope. Hubble's field of view was too small for the kind of sky-grabbing such a project would require. And securing time on HST was always a dicey proposition. Better to have a satellite telescope of one's own. So, true to the Berkeley Lab tradition, Perlmutter, his colleagues, and the US Department of Energy agreed that they should build one.

Over the next few years they drew up designs and, in a giant hangar in the hills above Berkeley, built cardboard models. And as technology improved, they rejiggered the designs and refashioned the models. They published glossy brochures and produced optimistic press releases. A couple of times they thought they might be one phone call away from final approval. But if the Supernova Acceleration Probe—SNAP—was going to get off the ground, it was going to have to do so with the help of NASA, and NASA was not in the habit of agreeing to a $600 million space mission on the say-so of the scientists who would most benefit from the launch.

In 2004, NASA convened a Science Definition Team to determine the viability of the project. An advocate for the Energy Department and an advocate for NASA would co-chair. For its own advocate, NASA called on Charles L. Bennett, a veteran of COBE and the principal investigator on the Wilkinson Microwave Anisotropy Probe, the satellite successor to COBE, named after former "Dickie bird" David Wilkinson, who died in 2002.

"But I don't know anything about dark energy!" Bennett protested.

Perfect, he was told. He would meet with dark-energy experts in complete ignorance and ask fundamental questions.

In the first meeting Bennett asked, "Is this going to be the last dark-energy mission? When you fly this mission, are you going to learn everything that you need to or can learn? Or is this just the first one, and then later you're going to have another?" Whether because, to someone who did know something about dark energy, the answer was obviously that you weren't going to learn everything about dark energy in one mission, or because, as Bennett suspected, no members of the SNAP team had ever asked themselves that question, there was silence. And it unnerved him.

In the end, he advised NASA to hold a "serious competition." Open it up to the community. See what other approaches might be out there. "Space," he said, "is not the best place to try a crapshoot."

In 2005 Bennett left NASA for Johns Hopkins. There he found himself in the office next to Adam Riess. Bennett told Riess he thought that Perlmutter might have "froze in too early" on supernovae. The choice of method made sense in 1999, but, as he had seen in his role as co-chair for the Science Definition Team, other methods had arisen since then. "Let's erase the blackboard and start from scratch," he said to Riess. They invited other collaborators, and they called themselves ADEPT, for the Advanced Dark Energy Physics Telescope—an acronym suggestive of astronomy's fast-and-loose image of itself, and a name notably lacking in what method the satellite would be using, because the collaborators themselves didn't know.

But the competition was also indicative of how dark-energy astronomy was changing. It had entered a new era. Carlton Pennypacker had likened the early search for distant supernovae to the Humphrey Bogart western *The Treasure of the Sierra Madre*, and although he hadn't survived the final reel, his gold dust had indeed led to a gold rush. Perlmutter and Riess were still around, working the mine, but they had plenty of competition—more and more prospectors bearing better and better equipment.

In the autumn of 1999 the National Research Council had initiated a study of what science could accomplish "at the intersection of astronomy and physics"—a topic that vindicated David Schramm's vision a quarter of a century earlier. Mike Turner chaired the committee, and he dedicated the resulting study to Schramm: *Connecting Quarks with the Cosmos: Eleven Science Questions for the New Century*. The first question was "What Is Dark Matter?" But the second question was "What Is the Nature of Dark Energy?"

Its *nature*. Not what it is, but what it's like. What it does. How it behaves. Like dark-matter astronomers, dark-energy astronomers had to confront a paradoxical question: How do you see something you can't see? And like dark-matter astronomers, they had to expand their understanding of "seeing" until it could encompass some manner of "coming into contact with." In doing so, however, they couldn't content themselves with the possibility of one day trapping a neutralino and hearing the ping, or converting an axion into a photon. Dark energy wasn't going to be a particle. The goal wasn't to detect it but to *define* it.

In particular, astronomers wanted to know if it was truly a cosmological constant — unchanging over space and time — or quintessence — something that did change over space and time. If it was unchanging, then as the universe expanded and the density of matter decreased, dark energy's influence would become greater and greater, leading to faster and faster acceleration, and the universe would indeed devolve into a Big Chill. If it changed over space and time, then it would be some kind of dynamical field previously unknown to physics, so it could, as far as anyone knew, either accelerate or decelerate cosmic expansion in the distant future. "In a Universe with dark energy," Michael Turner wrote in 2001, "the connection between geometry and destiny is severed." "What Is the Nature of Dark Energy?" might have rated second to "What Is Dark Matter?" in the *Connecting Quarks with the Cosmos* survey of science questions — and would perhaps have ranked number one if the topic weren't still so new when the committee finalized the list in January 2002 — but it was, the report said, "probably the most vexing."

That vexation led NASA, the NSF, and the US Energy Department to commission a task force. This time it was Turner's old partner-in-pizza, Rocky Kolb, who chaired the committee. The Dark Energy Task Force, which released the results of its deliberations in 2006, recommended four methods for investigating dark energy's nature.

One was the old standby, Type Ia supernovae. In the thirty years since Stirling Colgate devised his unsuccessful remote supernova search in the New Mexico desert, and Luis Alvarez challenged Rich Muller to see if he could succeed where Colgate had failed, astronomers had come down from the mountaintops. The two supernova teams from the 1990s had done most of their work in the control rooms of the telescopes — for instance, on Mauna Kea, the dormant volcano on Hawaii's Big Island where astronomers would sit at computers, muddy-headed from the 4,200-metre altitude. By the

the late 1990s, astronomers using those same telescopes were actually sitting in a control room at sea level, in an office on a commercial strip of shops in nearby Waimea. A few years later they were migrating to their own offices, back in Baltimore or Berkeley or Cambridge or La Serena or Paris. In the spring of 2007, the principal investigator of Berkeley Lab's Nearby Supernova Factory broke both his ankles in a fall at home and didn't miss a night of observing. He simply installed himself in a burgundy leather armchair in his living room, popped open his laptop, and monitored the University of Hawaii's 2.2-metre telescope, on Mauna Kea. Petting the dog with one hand while working the keyboard with the other, he spent his evenings studying lists of supernovae for follow-up, asking himself, "Which of these guys should I keep, and which should I throw away?"

So many supernovae, so little time.

Who wasn't working from a laptop in an airport? Who wasn't wireless? Who wasn't sending e-mail to the person next to him or her in bed? But if you had been around in the era when the detection of one distant supernova was a thrill without parallel, then having twelve or seventeen supernovae on your laptop was not something you took for granted.

Or having twelve or seventeen supernovae anywhere, laptop or not. Observing three nights a week for nine months of the year, the Nearby Supernova Factory was designed to discover about 150 or 200 supernovae annually, of which fifty or sixty would be Type Ia. And it wasn't the only factory out there. The Supernova Legacy Survey, a collaboration using the Canada-France-Hawaii Telescope on Mauna Kea, discovered 500 Type Ia during the decade. The Sloan Digital Sky Survey–II Supernova Survey discovered about another 500. The Center for Astrophysics Supernova Group—185. The Lick Observatory Supernova Survey—about 800. The Carnegie Supernova Project—around 100.

For some purposes, quantity was important. The premise of the Nearby Supernova Factory, for example, was that astronomers were never going to know the intrinsic brightness of a supernova. To perform the searches in the 1990s, they had invented methods to standardize supernovae. But in order to further refine their measurements, they needed a wealth of nearby supernovae in all sorts of varieties, in order to have a basis of comparison for whatever kinds of distant supernovae nature might throw their way.

Such as the distant supernovae that Adam Riess was pursuing with his own collaboration, now calling itself the Higher-z team. His showstopping presentation of SN 1997ff at the 2001 Dark Universe meeting offered persuasive evidence that at some point the universe had "turned over"—that the expansion went from decelerating to accelerating, from slowing down under the gravitational attraction of matter to speeding up under the countergravitational force of dark energy. Riess applied for more Hubble time to study supernovae, and in 2003 his team announced that they had determined *when* the universe turned over—about five billion years ago. In 2004 and 2006 his team produced evidence that even when dark energy was losing the tug of war with matter, as long as nine billion years ago, dark energy had nonetheless been present in the universe.

Another method the Dark Energy Task Force recommended was baryon acoustic oscillations, or BAO. In 1970 Jim Peebles had noted that in the creation of the CMB, the cosmological perturbations would have excited sound waves ("acoustic oscillations") that coursed through the primordial gas, creating peaks at intervals of 436,000 light-years. As the universe expanded, so did the spacing between these peaks; today they were 476 million light-years apart. And because galaxies tended to form on the peaks of these very large waves, astronomers could measure galaxy distribution at different eras, allowing them to see how the peak spacing changed over time—and thus how fast the universe had expanded over time. Whereas Type Ia supernovae behaved like standard candles, the spacing between the peaks acted like a standard ruler. But 476 million light-years was a lot of sky, even for cosmology. Astronomers needed enormous swaths of the whole universe just to lay the "ruler" on the map—technically impossible until 2005, when the Sloan Digital Sky Survey mapped the locations of 46,748 galaxies.

The third was weak lensing, the distortion of light from distant galaxies through the gravitational influence of foreground clusters of galaxies. Astronomers had been using this method to "weigh" dark matter by determining the shapes of millions of galaxies at various distances, which provided a direct probe of the mass of intervening clusters. After 1998, they began using weak lensing to measure the numbers of clusters over the evolution of the universe. That clustering rate depended on how fast the universe was expanding, and therefore the effects of dark energy, at different epochs.

The final approach—the one that Holzapfel was using at the South Pole Telescope—also used galaxy clusters. It was an effort to detect the Sunyaev-Zel'dovich (SZ) effect, named after the two Soviet physicists who predicted its existence in the 1960s. As a CMB photon made its journey from the primordial fireball to us, it might interact with the hot gas of a galaxy cluster, an encounter that would bump it up in energy—and out of the bandwidth the telescope was observing in. When that photon reached the SPT, landing on a microscopic thermometer at the heart of an ultra-cold (0.2 K) gold-plated spiderweb, it would appear as a hole in the CMB. There was something koan-like about the methodology: To see the unseeable, make the visible invisible.

"Very exciting!" Holzapfel said one afternoon, entering the Dark Sector lab that served as headquarters for the South Pole Telescope. Sitting at the controls was an incoming PhD student at Berkeley. She was knitting. "I can see the excitement is at a fever pitch," Holzapfel added.

She shrugged and said she would prefer to be handling Bakelite knobs and huge levers. But that wasn't how telescopes worked anymore. "I hit 'go' and wait twenty minutes for the script to run. At least this way"—she held up her knitting—"I get science and a sweater."

The United States established a year-round presence at the Pole in 1956, and the National Science Foundation's U.S. Antarctic Program long ago got everyday life there down to a science, as it were. Not that the South Pole wasn't still the South Pole. The case of Jerri Nielsen, the doctor who, during the austral winter of 1998, diagnosed her own breast cancer, took a biopsy, and administered chemotherapy, probably couldn't have happened anywhere else on Earth. And then there was Rodney Marks, an Australian astrophysicist who died suddenly on 12 May 2000; not until sunrise several months later could his body be flown to New Zealand for an autopsy, which revealed that the cause of death was methanol poisoning and raised the possibility that one of his fellow Polies had committed the perfect murder.

Usually, though, workers at the Pole referred to struggles for survival with irony, as if they, too, might encounter the kind of hardship faced by the crew fighting an alien-from-outer-space rampage in *The Thing*—not the 1952 original, which was set at the North Pole, but the 1982 John Carpenter remake, which took place inside the iconic geodesic dome that had served as science headquarters at the South Pole since the mid-1970s. In early 2008 a new base station officially opened, replacing the geodesic dome (which remained partly visible above the snow). But the new station resembled a small cruise ship more than a desolate outpost. It could house two hundred in private quarters. Through the portholes that lined the two floors, you could contemplate a horizon as hypnotically level as any ocean's. The new station rested on supports that, as snow accumulated over the decades, would allow it to be jacked up two full storeys. Amenities included a a state-of-the-art fitness centre, a gym, a twenty-four-hour cafeteria, a greenhouse, a computer lab, TV rooms with sofas deep enough to hibernate in, and Internet access about nine hours a day, when the communications satellites were above the horizon. As one mechanic said, looking out the cafeteria window at the New Year's Day revelers posing in swimsuits, beach-blanket style, at the ceremonial pole, "Hey, it's a harsh continent, haven't you heard?"

Creature comforts presuppose the existence of creatures, and you don't put creatures into an environment that supports no other creatures unless you've got a really good reason. The National Science Foundation thought it did: science you couldn't do anywhere else on Earth. Less than 800 metres from the South Pole Telescope, construction on the IceCube Neutrino Detector was redefining "telescope" by pointing its detectors not up at the sky but down through the earth. It was going to cover a square kilometre and consist of a series of eighty or so cables garlanded with sixty sensors each and descending (with the help of massive hot-water drills to clear the passage) about 1.5 kilometres below the snow. Those sensors should be able to observe the kinds of particles from space that can rip through the Earth's atmosphere, zip through the surface on the other side of the planet, and just keep going through crust, mantle, and core without interacting with anything else — unless, in a few cases, they smacked into an atom in the pure ice below the polar surface. (It wasn't a dark-matter experiment, but some of those particles might even be evidence of two dark-matter particles annihilating each other.)

In 1991, the NSF began a collaboration with the University of Chicago on the Center for Astrophysical Research in Antarctica. The purpose of CARA was to establish an observatory at the South Pole that would serve as a permanent base for millimetre and submillimetre astronomy — the Dark Sector, a tight cluster of telescopes, about a kilometre from the station, where light and other sources of electromagnetic radiation would be kept to a minimum. (Not far away were the Quiet Sector, for seismology research, and the Clean Air Sector, for climate projects.)

Holzapfel had been part of CARA almost from the start. He arrived at the University of Chicago as a postdoc in 1996, and although he returned to Berkeley as a faculty member in 1998, he continued to collaborate with CARA, and in particular with John Carlstrom, the director of CARA and the "dean" of Antarctic astronomy. Growing up in Pittsburgh, the son of an accountant and a teacher, Holzapfel developed an affinity for science for reasons he never understood. It was often science "as antisocial as science can be" — lots of high-voltage explosives and electrocution accidents. But it was also quiet, private science, like building a crystal radio and having a hard time believing that he was listening to chatter from the other side of the world.

And then, as part of CARA, he got to listen for whispers from the other side of the universe. "Very excitable," restless in the extreme, the size-15 trainer at the end of a crossed leg always vibrating as if agitating for more, Holzapfel helped conceive the experiments, design and run the instruments, and interpret the data from a series of CARA telescopes. Starting in the late 1990s, those experiments refined COBE's measurements even further season after season, right through the culminating project for CARA, the Degree Angular Scale Interferometer (DASI).

At first, DASI was no different (broadly speaking) from the others. It looked for patterns in the CMB — the temperature, the fluctuations — and found them. In April 2001, Carlstrom announced that DASI had indeed detected the telltale pattern of acoustic waves predicted by inflation; just as a musical note has overtones, the fetal cry of the universe should have three peaks.

The following year, however, DASI looked at the polarization—the direction of the photons as they decoupled from matter. The temperature and fluctuations told you where the matter was when the universe was 400,000 years old; the polarization told you how it was moving. Once again, the new cosmology faced a test. As the DASI team said in their PowerPoint presentations:

> if it's not there at the predicted level, we're
> back to the drawing board

It was there at the predicted level. No surprise, but a relief nonetheless.

And then came the Wilkinson Microwave Anisotropy Probe. In 2003 WMAP released its first set of data: another baby picture of the universe, a gentle riot of hot reds and cool blues representing the temperature variations that are the matter-and-energy equivalent of the universe's DNA. The match between simulations and data? Exact, only more so (if that were possible).

The South Pole Telescope was looking at the background radiation, too. But its mission wasn't just to do more of the same. It wasn't just documenting the radiation from the Big Bang in greater and greater detail, setting tighter and tighter margins of error for the next generation of CMB detector to beat. It wasn't using the CMB as an end in itself—a passive map, flattened on a celestial tabletop.

Instead, the SPT astronomers were using the CMB as a means—an active tool, one that would probe the evolution of the universe.

The construction of the South Pole Telescope required shipping 235 metric tons of material, first to Christchurch, New Zealand (the treaty-agreed port of entry to Antarctica), then to McMurdo Station, on the perimeter of the continent, and then, at the rate of 4,500 kilograms per LC-130, on twenty-five flights to the Pole itself. And because much of the technology at the Pole is singular, there were no economies of scale. If you were the PhD student who had to tighten a bolt during the construction phase, you couldn't just grab a precalibrated industry tool. You had to take off your gloves and use your hands, learning what one-sixteenth of a turn felt like.

Astronomers like to say that for more pristine observing conditions they would have to go into outer space—and Holzapfel thought of the people who wintered there as astronauts of a sort. Each year two grad students or postdocs took Pole duty on the SPT. Twice a day, six days a week, from February to November, the two "winter-overs" layered themselves with thermal underwear and outerwear, with fleece, flannel, double gloves, triple-thick socks, padded overalls, and puffy red parkas, mummifying themselves until they looked like twin Michelin Men. Then they trudged in darkness across the same plateau of snow and ice as the summer crew to look for the silhouette of the South Pole Telescope's 10-metre dish, except instead of trying to spy it through a whiteout, they identified it by how it blocked out a backdrop of more stars than any hands-in-pocket garden observer has ever seen. The telescope gathered data and sent it to the desktops of distant researchers; the two winter-overs spent their days working on the data, too, analysing it as if they were back home. But when the telescope hit a glitch and an alarm on their laptops sounded, they had to figure out what the problem was—fast.

They had to know what to do if—as happened once during the dark months—the instrument started making noises like a sledgehammer on steel: go outside, climb into the dish, and relubricate one of the bearings. Or a fan might break because the atmosphere was so arid that all the lubrication evaporated, and then the computer would overheat and turn itself off, and suddenly the system would be down, and nobody would have any idea why, and the telescope would be losing observing time at the rate of thousands of dollars an hour. And if the winter-overs couldn't fix whatever was broken, it would stay broken; planes don't fly to the Pole from February to October (the engine oil would gelatinize).

The job of summer crew members like Holzapfel was to prepare the instrument so that the "astronauts"—the winter-overs—didn't encounter any surprises during their six-month "spacewalk." Once a year the summer team would bring the SPT's detector in for a check-up—"in" being the control room below the telescope.

You could think of the SPT as a matryoshka doll. The outermost doll was a shield surrounding the antenna, to block as much light from the ground as possible. Next was the antenna, a 10-metre parabolic dish. Hovering above the dish, attached to a boom, was a "retractable boot," a long, rectangular metal container, which contained the receiver cabin, which held the receiver, which received the CMB photons through a window that opened on a secondary mirror, which bounced the photons toward six wedge-shaped wafers in a circle, like pizza slices, each wafer containing 160 bolometers, each bolometer containing a detector: the gold-plated spiderweb to catch CMB photons and, at the web's centre, a superconducting film 30 micrometres in diameter, or about half as thick as a human hair.

The grad-student-to-be put down her knitting and tapped her keyboard. To get at the innermost dolls, the team had to first turn the antenna to position the retractable boot over the roof of the laboratory building, then lower the boot until the receiver box nestled snugly over a matching panel in the roof of the control room. Inside the control room, the summer crew opened the ceiling and, using chains and brute force, extracted the receiver and gently guided it to the control room floor. After waiting thirty hours or so for the cryostat to warm up to room temperature, they pulled out their tools. At that point, one Pole veteran turned to another and said, "Do you know how many screws there are?"

"No. Hundreds. But the sad thing is, I've put each of them in three times."

By now, the telescope had been taking data for two seasons; shortly before heading south, Holzapfel had signed off on a paper reporting the serendipitous discovery of three galaxy clusters using the SZ method. In that method, the CMB provided a backlight of sorts on the foreground evolution of the universe. How the photons in the CMB had changed over the course of their journey through the universe would tell researchers how the universe itself had changed. The clusters that Holzapfel and his colleagues discovered by identifying this change—they were aiming for a thousand—would then undergo further scrutiny from other telescopes to determine their redshifts. When astronomers pieced together the abundances (determined from the SZ effect) and the distances (from redshift) of those clusters, they hoped to see the influence of dark energy on the growth of large-scale structure throughout the history of the universe—the same tug of war between dark energy and gravity that the other methods of defining dark energy were trying to detect. And that past was prologue. How the galaxy clusters had grown over the history of the universe would help astronomers predict which side would win that tug of war in the future.

Galaxy clusters were the largest gravitationally bound structures in the universe. Since gravity gathered smaller structures into larger ones, and gravity was now losing the tug of war with dark energy, it was reasonable to assume that galaxy clusters would also be the latest-forming gravitationally bound structures in the universe. And as dark energy took a greater and greater toll, they would also be the last-forming such structures.

Holzapfel thought of these clusters as the proverbial canaries in a coal mine. If the density of dark matter or the properties of dark energy were to change, the abundance of clusters would be the first thing to reflect that change. The South Pole Telescope should be able to track that change over time. At so many billion years ago, how many clusters were there? How many are there now? And then compare them to your predictions—your computer simulations—until they matched.

Holzapfel already had a hunch what that match would be. It was where all the methods of defining dark energy—the supernovae, the BAO, the weak lensing—were converging: the cosmological constant. He would have to abide by whatever the data said, but he didn't have to like it. He would prefer the other ending, the one where the universe collapses and then bounces back—the ending that speaks of rebirth and reminds us of seasons. Instead, the story of the universe appeared to be heading toward the conclusion you could see, metaphorically, everywhere you turned at the South Pole: cold and empty and eternal: the clusters receding until we won't see them anymore. A hundred billion years from now we'll be left with just one cluster, our Local Group, and no clue that anything else is out there.

Like many astronomers, Holzapfel found that outcome "depressing." Not for some woe-is-me reason; he already considered himself "existentially challenged." He was perfectly content to liken life to a Russian novel, in which a depressing future can be as exciting as happily-ever-after. His concern was more professional. He didn't like a story of the universe that ended with his profession — cosmology — dying out.

But then, nobody ever said the universe had to be benign.

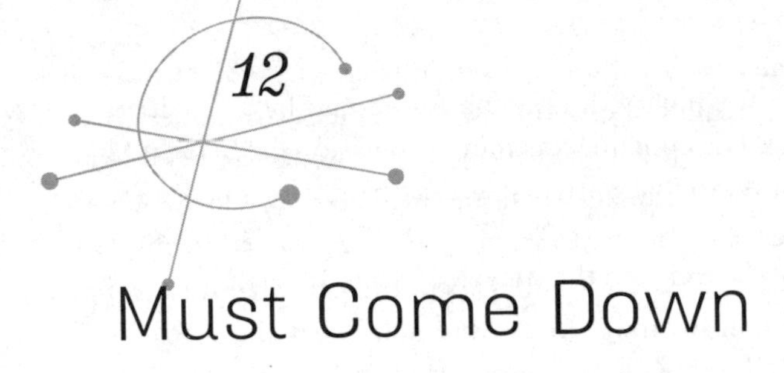


# Must Come Down

THEY NEEDED SOMETHING to write on—now. The discussion had progressed to the point where words wouldn't do. They needed numbers, signs, the propulsive force of mathematical symbols flying across a surface. The table of theorists got up and joined the several other clutches of theorists at work on the only blackboard in the room. Still, there was plenty of space. The blackboard was "full wall," as they liked to say at the Perimeter Institute for Theoretical Physics Physics. Blackboards in offices were full wall. Blackboards in the hallways, blackboards in nooks off the hallways, blackboards in outdoor courtyards—all full wall. The blackboard in the café reached floor to ceiling, and stretched the length of the room. The theorists had all turned their backs on the café tables, on the windows, on the view of the sunset. Here, there, along the wall, they hunched forward, peering at the hieroglyphs appearing on the board, gesturing their concerns, voicing their corrections. The new group, however, had no chalk. No matter. They simply bent close to the blackboard and waved their hands, their fingers describing arcs in the air. They didn't need chalk. For them, the equations were *there.*

From across the café, Brian Schmidt watched. "They're really going at it," he said to nobody in particular. Then he produced a mobile phone and took a picture.

As the leader of the original High-z team, Schmidt was one of the dark-energy astronomers whose discovery nine years earlier had sent physicists down the byzantine path leading to this blackboard. Now he had entered the theorists' den. When the weeklong meeting on dark energy began, with a four-day conference at McMaster University, in nearby Hamilton, Ontario, several other astronomers had been in attendance. But today the setting had shifted seventy kilometres northwest to the Perimeter Institute, in Waterloo, Ontario, and the number of participants had thinned considerably. "I'm the last astronomer standing," Schmidt had said to the organizer of the Perimeter event, who answered, "No, you're not. What about Rocky?" Schmidt laughed. Rocky Kolb was as much an astronomer as Schmidt was a theorist, and Schmidt had made a point, in his lecture a day earlier, of identifying himself as a "dyed-in-the-wool astronomer." In the mid-1990s, when Schmidt had delegated the responsibilities for the suite of publications that would present the High-z results, a theorist had told him he would need to include a paper on something called the equation of state; Schmidt had shrugged, said "Okay," and invited his former Harvard officemate Sean Carroll to advise on the topic. Back then Schmidt hadn't even known what the term "equation of state" meant; now it seemed to be all anybody wanted to talk about, not just at the conference this week—in May 2007—but at every other dark-energy conference.

Cosmology had a new number. Just as omega quantified the density of mass, the equation of state quantified the density of energy—specifically, the ratio of pressure to energy density. Cosmologists designated it as *w*. A cosmological constant would mean that *w* was exactly equal to –1; Einstein's lambda proposes that a given volume of space should have an inherent amount of energy per unit of volume, and that this energy suffuses the universe and remains constant over time. A *w* not –1 was quintessence. It would do . . . something else.

As the tenth anniversary of the discovery approached, the number of dark-energy meetings was only growing. As a resident of Australia, Schmidt had to travel halfway around the world just to get anywhere; at dinner a couple of nights earlier, he'd joked to his colleagues, "My average velocity for a year is 70 to 80 kilometres per hour." But the setting almost didn't matter. For the participants, the meetings and the message were becoming stupefyingly similar: the same chorus delivering variations on the same theme—a Mike Turner or a Saul Perlmutter here, a Rocky Kolb or an Adam Riess there, all of them looking for answers and coming up empty: a movable famine.

But, if you had to attend a conference, the Perimeter Institute at least had a marble-top bar where if you ordered a glass of wine, the bartender produced a wine list. Schmidt—who, in the thirteen years since he'd helped form the High-z collaboration in 1994, had graduated from scrappy postdoc to vineyard owner—approved. The Perimeter Institute began life in 2000 with a £70 million endowment from Mike Lazaridis, the founder of Research in Motion, which created the BlackBerry. The guiding principle was, as at many institutes, to give theorists a place to think free of distractions. The difference with Perimeter was that the freedom came with luxury. The interior design of the building alternated between full-wall windows and exposed concrete. A four-story atrium divided administrators from theorists. For lunch, the theorists could stop by the café, or they could stay in their office and order room service.

The first part of the week, at McMaster, had consisted of the usual conference-style presentations: one talk after another in an auditorium seating a hundred or so participants. The Perimeter part of the week, however, would be a workshop: talks open to interruptions, catch-as-catch-can discussions in smaller groups, and the chance to keep on talking in the hallways and in the nooks and on the terrace and, always, over snacks and coffee and meals in the Black Hole Café. At dinner on the first evening at Perimeter, astronomer Schmidt and theorist Christof Wetterich, of Heidelberg University, fell into a discussion about a distinction Kolb had made earlier that afternoon, during the final lecture at McMaster.

Kolb had begun with a meditation on how scientists think about cosmological models. To the astronomers of Copernicus's day (though not necessarily to Copernicus himself), a cosmological model was a representation of a world that made mathematical sense but might bear no relation to reality. Whether the Sun or the Earth was at the centre of the cosmos didn't matter; what mattered was which object was more mathematically useful at the centre of the cosmological model. Not so for the scientist of today. Over the past four centuries, scientists had learned that the accumulation of evidence could tell them which model was more correct—the one with the Earth at the centre or the one with the Sun at the centre. Today astronomers regarded the creation of a cosmological model as an attempt to capture "reality itself," Kolb said. "We really think that dark matter is a reality, and that dark energy is a reality." If they somehow turned out not to be, fine. But "that's really what we have to test."

Sitting in the Black Hole Café, Schmidt took a sip of wine, pronounced it palatable, and said that he had to disagree with Rocky. "There is no reality," he said to Wetterich. "There are only predictions."

Wetterich said he had to agree with Rocky. To make his point, he picked up a water glass. "If I drop it," he said, "it will fall to the table."

Schmidt shook his head. Yes, he conceded, every glass throughout history, when released, has fallen. "But once it might not," he said. "You can only *predict* that it will—to a high degree of confidence," he added.

"I believe it will," Wetterich said.

Schmidt shrugged. "I hope it won't."

The meaning of reality might seem a subject best left to philosophers, but, like philosophy itself, it had always been the domain of physicists, too. The ancients thought they couldn't capture "reality," so they settled for saving the appearances. Once Galileo had provided empirical evidence that Copernicus's Sun-centred system was correct, and once Newton had codified the maths, scientists came to understand that equations on paper could do more than approximate reality: If you could find it in the heavens, you could capture it on paper—the point Kolb was trying to make. Then Einstein came along and reversed that process. If you could write it on paper, you could find it in the heavens.* If your equations told you that time passed differently for two observers moving in relation to each other, or that gravity bent light, then that was what nature did. You would, Einstein acknowledged, have to test those predictions: "Experience remains, of course, the sole criterion of the physical utility of a mathematical construction"—the point Schmidt was making to Wetterich. If you found an exception in nature, then you either adjusted or abandoned the theory. But Einstein, speaking from his own experience, then proceeded to argue the point that Wetterich was making to

* More precisely, Einstein argued that this was the logic physicists had already been following; they just didn't know it or, in some cases, including that of Einstein early in his career, refused to acknowledge it.

Schmidt: "I hold it that pure thought can grasp reality, as the ancients dreamed."

Schmidt and Wetterich weren't going to settle the debate. It was ancient; it was eternal. For Schmidt, though, it was also personal. In 1998, he'd let go of a glass and it went up.

However transcendent the discovery, the human repurcussions sometimes weighed heavily on Schmidt. This was one of those times. That spring, just prior to the McMaster and Perimeter meetings, Schmidt had heard from the Peter and Patricia Gruber Foundation—a philanthropy made possible by a Wall Street fortune—that he and Perlmutter were the recipients of that year's Gruber Prize in Cosmology, worth about £250,000. They were in excellent company. Jim Peebles and Allan Sandage had shared the first cosmology prize, in 2000; other recipients had included Vera Rubin in 2002, and Alan Guth and Andrei Linde in 2004. The apportionment of credit for the 2007 prize was understandable. According to the unwritten rules of science, Brian Schmidt was the big gun on the High-z team. But the team had deliberately tried to rewrite those unwritten rules, and Schmidt was still trying: All week he had been negotiating with the foundation, requesting that Adam Riess be added to the list of recipients because he had been the author of the "discovery of acceleration" paper for the High-z team.

The nuances were more than academic. Once while Perlmutter was making a presentation at a conference, Nick Suntzeff turned to Bob Kirshner and whispered, "Saul thinks there's a Nobel Prize in this."

Kirshner gave Suntzeff a look. "There is!"*

The Grubers were, in a way, the Golden Globes to the Nobel's Academy Awards. The preceding year, 2006, the Gruber Prize in Cosmology had gone to John Mather and the COBE team. Then a few months later the Nobel Prize in Physics went to Mather and his COBE collaborator George Smoot. That the discovery of acceleration was worthy of a Nobel Prize wasn't subject to much debate. What was debated was who made the discovery. "Saul is going to win a Nobel Prize," Alex Filippenko would say, with a shrug. "My only

---

* Or "N**** Prize," as some scientists, including Kirshner, often prefer to write it; apparently science can banish medieval superstition only so much.

hope is that the Nobel committee will do the right thing and give it to Brian and Adam as well. That would be the fair thing. *All right?*"—this last said as if there were someone challenging him, which in a sense there was.

The two teams had long before informally agreed that the discovery was, as Riess would say, "big enough and cool enough to share." The standard construction was that in early 1998 two teams had independently reached the same surprising conclusion—that the expansion of the universe appeared to be accelerating. In June 2006, Perlmutter, Riess, and Schmidt learned that they had won the Shaw Prize in Astronomy, a £500,000 award endowed by the Hong Kong media magnate Sir Run Run Shaw in 2002. (Jim Peebles had gotten here first, too, in 2004.) So far, so good. But the following month, July 2006, came the announcement that the Antonio Feltrinelli International Prize in Physical and Mathematical Sciences—awarded once every five years by Italy's Accademia Nazionale dei Lince, or Academy of Lynxes, dating back (albeit with a centuries-long interruption) to Galileo's day, and carrying an award of about £170,000—was going to Perlmutter . . . and only Perlmutter.

Members of the High-z team interpreted this news as evidence that the Berkeley Lab "publicity machine" had done its work. They still recalled how George Smoot in 1992 had broken the agreement not to publicize the COBE results before the public announcement; to make matters worse, as Mather wrote in a book about the project, the LBL press release "mentioned NASA only in passing and did not cite a single member of the COBE science working group other than George." Now, members of the High-z team feared, the LBL press office was performing the same beyond-the-call-of-duty (and possibly of ethics) service for Saul, making him "appear to be God and the greatest thing since sliced bread," in the words of Filippenko. It had persuaded someone, somewhere, in a position of influence that in the 1 January, 1998, *Nature* paper, or maybe at the 8 January 1998, AAS press conference, or maybe among the 9 January 1998, AAS posters, the Supernova Cosmology Project had announced the discovery of cosmic acceleration.

Members of the SCP, however, had long thought that the High-z team was trying to discredit *them*. In 2001 Filippenko's personal account of his experience as the only astronomer who was a member of both teams appeared as an article in *Publications of the Astronomical Society of the Pacific;* the following year Bob Kirshner published his own personal account as a book, *The Extravagant Universe.* After Mike Turner reviewed Kirshner's book for *Science,* LBL's Robert Cahn — still seething from his own experience having to defend the SCP from Kirshner's recommendation to shut it down, followed by Kirshner's appropriation of Perlmutter's Hubble Space Telescope plans — said to Turner, "Well, it tells me something that after their results, Saul set out to design an experiment in space to really understand this, and Kirshner decided to write his memoirs." Or as one member of the SCP collaboration remarked, basically summarizing the response of the whole team, "The High-z accounts don't recognize the other side of the Mississippi." For his part Perlmutter asked the LBL publicity department to compile a preemptive history of the discovery, then file it away for the day it might be most useful.

As the anniversary of the discovery grew closer, tensions grew stronger. At one cosmology conference "celebration" Gerson Goldhaber reviewed the history of the discovery, concentrating on his histogram from the autumn of 1997. "I'm mentioning the dates," he said, "because the date of discovery is of some importance." He also referenced the colloquia that Perlmutter had given late that year. "Now, the question is, who remembers all that?" he said. "Well, Saul's talk was videotaped, and I have the videotape."

Riess followed him. He opened his speech by wishing Goldhaber a happy birthday. Goldhaber nodded and accepted the applause of the audience. Then Riess showed some of his team's e-mails from January 1998. And then he showed a clip from his TV appearance on the day that the journal *Science* published the article on Filippenko's talk at UCLA.

The following morning it was Perlmutter's turn. He opened by saying he hadn't planned to focus on the past, but he did want to show that on 9 January 1998, his team had their "equivalent" media moment with the publication of the front-page article in the *San Francisco Chronicle* covering his participation in the AAS press conference, and although he didn't have e-mails from that period, he did have the minutes of the team's meetings from the autumn of 1997, which he also displayed.

"Dear Saul," Bob Kirshner began a letter dated 12 January 2007:

> I was dreamily thinking about 2006, as people often do on the final day of the year.
>
> As part of this idle foolishness, I took a look at the Shaw Prize site. I want to say again that I think this award is a great thing and I am very glad the cosmic acceleration is being recognized. You have a lot to be proud of, and I feel the same way about the work that Adam and Brian and the rest of us have done.
>
> But there was one point that jumped out at me and I can't get it out of my mind. That is why I am writing to you.

Over the years Brian Schmidt had come to recognize that Kirshner's promotion of Schmidt's thesis work at conferences in the early 1990s—however much the two of them might have disagreed about the apportionment of credit—was part of what a big gun did: get word out to the community about an acolyte's accomplishments. Now Kirshner had taken on similar duties regarding the High-z team's accomplishments, peppering magazines and newspapers with letters and e-mails objecting to coverage that, in his opinion, favoured the SCP.

In this case, he was objecting to the mini-autobiography that Perlmutter had written for the Shaw Prize website, in particular this passage: "We announced these results at the American Astronomical Society January 1998 meeting. Because both our team and Brian's team—including Shaw co-winner Adam Riess—independently announced matching results at conferences in the beginning of the year, by the end of the year most of the scientific community had accepted the startling findings." Over seven single-spaced pages, Kirshner proceeded to quote from (and link to) the LBL's 8 January 1998, press release, contemporaneous press accounts, and books to support his conclusion that Perlmutter "**did not announce**," "**did not announce**," "**did not announce**," and, for good measure, "**did not** make an announcement" (boldface his) of acceleration in January 1998.

When Kirshner didn't hear back from Perlmutter, he revised the letter, removing the salutation and the references to "you," among other modifications, and on 27 February 2007, he posted it on his Harvard website under the link "Thoughts on the discovery of dark energy."

"Dear Bob," Perlmutter finally replied, in a letter dated 12 June 2007:

> Now that the teaching semester is over let me address the 9-page letter that you sent concerning our January 1998 AAS scientific presentation and press conference. As I mentioned in my earlier email I was greatly surprised by your letter, and in fact had previously been thinking I should email a request to a few members of the original High-Z team that they stop referring to our January announcement as "weaker" or "more tentative" than the High-Z team Marina Del Rey announcement, since I think this is incorrect. However, before your email I had never heard the suggestion made that we had not presented *any* substantive results at the January meeting. (Obviously, there is no question about which group's paper got out first, but you are clearly making a broader claim here that I believe misrepresents the history.)

The controversy was coming down to the meaning of the word "announce." A month later, responding to Perlmutter's letter, Kirshner addressed the issue directly: "Did you or did you not 'announce' the accelerating universe at the AAS meeting in January 1998? 'Announcing' is what you claim to have done in your Shaw autobiography. Twice. That's what your letter aims to show. After reading your letter, I am even more convinced that this is not correct." And he went on, for yet another seven single-spaced pages, to cite some of the same press accounts as before, as well as to rebut the references that Perlmutter had included in his June letter, including the front-page article in the *San Francisco Chronicle* that ran under the sub-headline "Universe getting bigger and bigger, faster and faster—forever," and reported that the SCP study "seems to indicate" that the "expansion is starting to speed up." *

* For his part, the article's author recalled sitting at his desk the day his piece ran and looking at other newspapers and media outlets to see how they had covered the big news, finding nothing, and thinking, "*What* the *fuck*."

This time Perlmutter didn't bother to respond.

During the pre-anniversary period, tensions grew stronger within the teams as well. In 2007, even while he was exchanging gritted-teeth pleasantries with Perlmutter, Kirshner managed to alienate some of his collaborators. For a talk entitled "Supernovae and the Accelerating Universe" at the Aspen Center for Physics, Kirshner presented "A Timeline of Important Developments" that, among multiple references to the High-z group, included exactly one mention of Brian Schmidt—a (near) omission that only reinforced the feeling among some team members that Kirshner took "too much credit for himself." Nor was the SCP team immune from this self-cannibalization. Gerson Goldhaber began circulating his own history of the discovery, which underwent several revisions to accommodate the complaints of some SCP collaborators who felt he was claiming the discovery for himself. Privately, though, Goldhaber stuck to his story: "My team found it first, and I found it for my team."

As part of the anniversary preparations, STScI—Reiss's home turf—sponsored a media day; Perlmutter flew in from the West the West Coast for the occasion. A week later, *Newsweek* ran an article on dark energy that led with a re-creation of Riess's 1997 calculation of a universe with negative mass, and went on to quote Kirshner on the cosmological constant. Not only did Perlmutter and the SCP receive no mention, neither did the existence of any discoverer other than Riess (aside from a reference to "and his colleagues").

That did it. Perlmutter contacted the LBL press office: The time had come to release the SCP version of the history of the discovery of dark energy. Shortly thereafter it appeared on the Berkeley Lab website as a three-part series. Part One began: "Saul Perlmutter, leader of the international Supernova Cosmology Project (SCP) based at Berkeley Lab, made the first public announcement of evidence for the accelerating expansion of the universe on January 8, 1998. . . ."

In an office just up the road from Harvard Square, a quivering hand reached for a keyboard.

This wasn't the legacy Schmidt or any of the other members of the two teams wanted for themselves: bickering boffins. And now they couldn't even guarantee that they had done their discipline proud.

"Fundamentalist Physics: Why Dark Energy Is Bad for Astronomy." The title alone would have guaranteed that the paper would get attention. That its author was Simon White, one of the directors at the Max Planck Institute for Astrophysics, in Germany, guaranteed that it would get serious consideration. That it presented arguments the community had begun finding unavoidable made it a sensation.

The paper appeared online in April 2007, in advance of publication in the journal *Reports on Progress in Physics* and just prior to the McMaster conference and Perimeter workshop. Like the contentiousness over the Gruber Prize, the paper preoccupied Schmidt that week, not only because everybody else was talking about it, but because he was sympathetic to a lot of what White had to say. The core of White's argument was that astronomy and particle physics constituted two different cultures. Astronomers, White said, were "generalists," exploring the complexities of the universe on a case-by-case basis. Particle physicists were "fundamentalists," wringing the complexities of the universe in the hope of squeezing out an "ultimate foundation"—a "Truth." "Dark Energy," he wrote, "is a unique link between them, reflecting deep aspects of the Fundamental Theory, yet apparently accessible only through astronomical observation."

In the theory-and-observation, call-and-response system of investigating nature that scientists had refined over the previous four hundred years, the dark side of the universe represented an irruption. Copernicus's heliocentric theory anticipated Galileo's observations of Jupiter and Venus, which inspired Newton's theory of universal gravitation, which anticipated more than two centuries' worth of moons, planets, and stars, which inspired Einstein's theory of general relativity, which anticipated the observations of the expanding universe, which inspired the Big Bang theory, which anticipated the observations of the cosmic microwave background, which inspired the revival of Einstein's theoretical cosmological constant, which anticipated the observations of Type Ia supernovae, which inspired . . . what? Not a theory, exactly. Just a name for a theory—and not even a theory. A theory-to-be: dark energy.

"We're desperate for your help," Schmidt had called to the theorists in the audience at another cosmology meeting a couple of years earlier. "You tell us what you need, we'll go out and get it for you."

To which the most succinct response was one his old officemate theorist Sean Carroll offered at yet another cosmology meeting: "We have *not a clue.*"

Not a clue, yet no end of ideas. Every day Adam Riess checked an Internet site where scientists posted papers; he was hoping for the paper that would finally present a "deep theory," but he found most of them "pretty kooky." Saul Perlmutter liked to begin public talks with a PowerPoint illustration: papers on dark energy piling up, one on top of the next, until the on-screen stack ascended into the dozens. Schmidt had looked online at how many papers cited the original dark-energy papers and found three thousand—of which 2,500 were theories. In his talk at McMaster, he had included a list of prospective candidates for dark energy that a friend had culled from the recent literature:

> Tracker Quintessence, single exp Quintessence, double exp Quintessence, Pseudo-Nambo-Goldstone Boson Quintessence, Holographic dark energy, cosmic strings, cosmic domain walls, axion-photon coupling, phantom dark energy, Cardassian model, brane cosmology (extra-dimension), Van Der Waals Quintessence, Dilaton, Generalized Chaplygin gas, Quintessential inflation, Unified Dark matter and Dark energy, superhorizon perturbations, Undulant Univese, various numerology, Quiessence, general oscillatory models, Milne-Born-Infeld model, k-essence, chameleon, k-chameleon, f(R) gravity, perfect fluid dark energy, adiabatic matter creation, varying G etc, scalar-tensor gravity, double scalar field, scalar+spinor, Quintom model, SO(1,1) scalar field, five-dimensional Ricci flat Bouncing cosmology, scaling dark energy, radion, DGP gravity, Gauss-Bonnet gravity, tachyons, power-law expansion, Phantom k-essence, vector dark energy, Dilatonic ghost condensate dark energy, Quintessential Maldacena-Maoz dark energy, superquintessence, vacuum-driven metamorphosis

"Time to get serious." The PowerPoint slide, teal letters popping off a black background, stared back at a roomful of cosmologists at yet one more conference. Sean Carroll had taken it upon himself to give his fellow theorists their marching orders. The "heyday for talking out all sorts of crazy ideas," as Carroll explained, was over—that heady, post-1998 period when Michael Turner might stand up at a conference and call for "irrational exuberance." Now had come the metaphorical morning after.

The observers had finished their work. They had used supernovae, weak lensing, BAO, galaxy clusters, and the cosmic microwave background to find more and more evidence for acceleration until the community agreed: The effect was genuine. Then the observers continued their work, trying to figure out if dark energy is quintessence or the cosmological constant. And they continued their work. And continued. "Dark Energy is the Pied Piper's pipe," White wrote "luring astronomers away from their home territory to follow high-energy physicists down the path to professional extinction."

For more than twenty years, particle physics had been pursuing one prey: the Higgs boson, a hypothetical particle that would explain the presence of mass in the universe. The Tevatron at Fermilab had been trying to create it; the Large Hadron Collider in Geneva was now trying to manufacture it. By the time the discovery of cosmic acceleration was ten years old, astronomers were beginning to wonder if they, too, were practicing a science in search of one result: *w*.

Schmidt recognized that he had done his part to usher astronomy into the arena of Big Science. Together the two supernova search teams had relied on the efforts of more than fifty collaborators. But the changes that astronomy was experiencing would have been happening anyway. Not only were the areas of study becoming more specialized — supernovae, CMB, gravitational lensing, and on and on — but so were the means of studying them, the narrow bands of the electromagnetic spectrum. The South Pole Telescope, for instance: the Sunyaev-Zel'dovich effect required astronomers to use a specific sub-millimetre wavelength in order to detect the "holes" left by photons that had moved out of that frequency. In one generation astronomy had gone from the lone observer on a mountaintop taking photographs in visible light to dozens of collaborators around the globe pursuing a variety of specializations by looking at increasingly narrow bands along the electromagnetic spectrum. Even if the supply of funding globally stayed the same, the demand for it wouldn't. Increasing specialization of areas of research and increasing diversification of research methods were creating an effective shortage of resources.

And nothing was demanding those resources like dark-energy research. ADEPT—the space telescope that Chuck Bennett and Adam Riess had conceived in response to Perlmutter's SNAP—had eventually adopted baryon acoustic oscillations as its primary observing strategy, though it would also incorporate Type Ia supernovae. SNAP had expanded its mission from Type Ia supernovae alone to weak lensing. The cost for either satellite would be at least £600 million—except NASA budgeted only $600 million. "It's not worth it!" one astronomer pleaded with a NASA representative at a dark-energy conference. "Either you do it right, or you don't do it. And if you're not going to do it right, then give us the money back and we'll do other things."

"Hear, hear," piped up Mike Turner from a seat near the front of the auditorium.

Schmidt was married to an economist, and he took an unapologetically bottom-line approach to the idea of a space telescope dedicated to dark energy: Was the mission worth the expense? What were the trade-offs? How much good science wasn't going to be done because of the community's concentration on dark energy? The answers would have been easier if the evidence so far were indicating that dark energy was not only quintessence but unmistakably quintessence—"standing out like dog's balls," as Schmidt would say, adopting the Australian vernacular. Studying something that creates a lot of changes in the cosmos, even subtly, would be more challenging, more satisfying, and probably more revealing of the universe's secrets than studying something that stays the same. If the magic number for omega had been 1, then the un-magic number for the equation of state was –1 because, as Jim Peebles said, "then it's a number, and we have nothing to do."

But as the first decade of dark energy drew to a close, the evidence seemed to be pointing to the cosmological constant. The more that observers continued to do their job, the closer they got to –1. The question then became, as the title of a session at one of the dark-energy conferences put it, "How Well Do We Need to Do?"—as in, "How close do we need to get to $w$ equals –1 before we agree that $w$ does equal –1?" Lawrence Krauss, a theorist and panelist for the session, laid out his argument thus:

The most reasonable theoretical prediction is $w = -1$.
Observations suggest $w = -1$.
Measuring $w$ approximately $= -1$ therefore tells us nothing.

So observers shouldn't pursue *w* at all, right? Wrong. "How well do we need to do?" Krauss said, repeating the title of the session. "Better than we'll be able to do! We need to do better than anything you're ever going to be able to do in the lifetime of the people in this room, I expect, experimentally. And that's just life. In spite of the fact that you're likely to spend the rest of your lives measuring stuff that won't tell us what we want to know, you should keep doing it. But you should be prepared to have a standard model for twenty or thirty years that you don't understand." Speaking on behalf of the theorists, he was basically telling observers, *Keep on doing your job; we'll catch up.*

Until they did (assuming they did!), science was stuck with—to invoke a term that Carroll helped popularize through articles, lectures, papers, and a blog—a "preposterous universe." It was a universe that had the benefit of a seemingly perfect match between observation and theory—the workings of the heavens and the equations on paper. Take the observations of supernovae and the cosmic microwave background, apply the theory of general relativity, and you had a universe that did indeed add up to the magic omega number of 1. But it was also a universe that didn't add up. Take the observations of supernovae and the cosmic microwave background, apply the other cornerstone of twentieth-century physics, quantum theory, and you got gibberish—an answer that was 120 orders of magnitude off.

Which didn't mean that dark matter and dark energy were the modern equivalents of epicycles or the ether. But it did mean that theorists had to confront the same problem that had consumed Einstein for the final three decades of his life: how to reconcile the physics of the very big—general relativity—with the physics of the very small—quantum mechanics.

Theorists could use the two theories simultaneously—for instance, Hawking radiation, Stephen Hawking's idea that while quantum mechanics dictated the existence of pairs of virtual particles at a black hole's horizon, general relativity dictated that sometimes one of those pairs would slip over the edge into the black hole and the other would rebound back into "our" universe. But theorists hadn't yet figured out a way to make the two theories work together—to make an observation of 0.7 consistent with a prediction of $10^{120}$. What made the two theories incompatible—where the physics broke down—was the foundation of the past four centuries' worth of physics: gravity itself.

In physics, gravity is the ur-inference. Even Newton admitted that he was making it up as he went along. In one of his letters to Richard Bentley about the stability of a universe operating under the influence of gravity, Newton wrote that the notion of a force of attraction existing between two distant objects is "so great an Absurdity that I believe no Man who has in philosophical Matters a competent Faculty of thinking can ever fall into it." Nearly two centuries later, the German philosopher-scientist Ernst Mach wrote, "The Newtonian theory of gravitation, on its appearance, disturbed almost all investigators of nature because it was founded on an uncommon unintelligibility." Now, he went on, "it has become *common* unintelligibility." Einstein endowed gravity with intelligibility by defining it not as some mysterious force between two objects but as a property of space itself, and he refined Newton's equations so that the presence of matter and the geometry of space are interdependent. Most of the interpretations of dark matter and dark energy arose from the right-hand side of the equation for general relativity, the side where Einstein put matter and energy. But there are two sides to every equation, and in this case what was on the other side was gravity.

Most astronomers had dismissed Modified Newtonian Gravity, or MOND, as an explanation for dark matter back in the early 1980s; after the 2006 release of the Bullet Cluster data—the photograph that "showed" dark matter separating out from regular matter in the collision of two galaxy clusters—even its defenders began to distance themselves. The Bullet data could accommodate MOND, but you would still need some kind of dark matter to explain the rest. And at that point MOND began to lose the attraction of simplicity.

But just because MOND might not be valid didn't mean that general relativity was. Back in the 1950s Bob Dicke had organized the Gravity Group at Princeton in part to put Einstein to the test, and his efforts helped inaugurate a generation of experiments. The discovery of evidence for such general-relativistic phenomena as black holes, pulsars, and gravitational lensing had only accelerated those efforts. Dark matter and dark energy, however, endowed those efforts with a sense of urgency.

Physicists were testing gravity on the scale of the very large. In the Sacramento Mountains of New Mexico, the Apache Point Observatory Lunar Laser-ranging Operation (APOLLO) aimed a pulse at the Moon twenty times every second. If a cloud happened to pass, then a green dot appeared, Bat Signal–like, in the purpling twilight over Alamogordo. Otherwise the beam traced a clear path to a target 383,000 kilometres away: one of three suitcase-size mirrors that Apollo astronauts had planted on the lunar surface four decades earlier specifically to facilitate this kind of experiment. With every laser burst, a few of the photons from the beam would bounce off the reflecting surface and complete their return journey to New Mexico. Total round-trip travel time: 2.5 seconds, more or less.

That "more or less" made all the difference. By timing the speed-of-light journey, researchers were measuring the Earth-Moon distance moment to moment and mapping the Moon's orbit with exquisite precision. As in the apocryphal story of Galileo dropping balls from the Leaning Tower in Pisa to test the universality of free fall, APOLLO treated the Earth and the Moon as two balls dropping in the gravitational field of the Sun. If the orbit of the Moon exhibited even the slightest deviation from Einstein's predictions, scientists might have to rethink those equations.

Physicists were also testing gravity on the scale of the very small. Until the turn of the twenty-first century researchers didn't have the technology to measure gravity at ranges of less than a millimetre, in part because testing gravity isn't simply a matter of putting two objects close to each other and quantifying the attraction between them. All sorts of other things may be exerting a gravitational influence. In a series of jewel-box experiments at the University of Washington, researchers had to take into account the metal of nearby instruments, the soil on the other side of the concrete wall that encircled the laboratory, the changing water level in the soil after a rainfall (and since the experiments were taking place in Seattle, that was a lot of changing), a nearby lake, the rotation of the Earth, the position of the Sun, the dark matter at the heart of our galaxy. Nonetheless, they narrowed their measurements of gravitational attraction down to a distance of 56 microns, or 1/20th of a millimetre.

So far, Einstein was holding, both across the universe and across the tabletop. And with every narrowing of the range where Einstein might have been wrong, another hypothesis died, and Brian Schmidt's PowerPoint slide lost one more esoteric name. Even if Einstein didn't hold, researchers would first have to eliminate other possibilities, such as an error in the measure of the mass of the Moon or the Sun, or in the level of groundwater after a spring shower, before conceding that general relativity required a corrective.

Even so, astronomers knew that they took gravity for granted at their own peril. Why was it so weak? Why—in the example that scientists commonly cited—could the gravitational pull of the entire Earth on a paper clip be counteracted with a pound-shop magnet?

Because, some theorists proposed, gravity was a relic from a parallel universe. Theorists commonly called these universes "branes," as in membranes. If two branes were close enough to each other, or even occupying the same space, then they might interact through gravity. Gravity would be on the scale of the other three forces if we had access to those other universes. What was a powerful force in a parallel universe might be the source of the effects of dark matter or dark energy in ours. The problem with these theories, at least from an astronomer's point of view, was how to test them. A theory needs testable predictions or it's not a truly scientific theory; its validity must come down to observations. But how can you observe a universe beyond your own?

Scientists are always wincingly aware that they are prisoners of their perceptions. It was true, for example, that if dark energy was the cosmological constant, then a hundred billion years from now all that cosmologists would see would be a handful of galaxies. But it was also true that we didn't need to wait a hundred billion years to confront a similar perceptual obstacle. Inflation already ensured that certain traces of the universe's initial conditions would be forever out of reach.

Those conditions would put even branes to shame, in terms of defying perceptions. If inflation can pop one quantum universe into existence, why not many? In fact, according to quantum theory, it should. It *would*, if inflation actually happened. In that case, our inflationary bubble would be one of an ensemble of $10^{500}$ inflationary bubbles, each its own universe. That's 100,000,000,000,000,000,000, 000,000,000,000,000,000,000,000,000,000,000,000,000, 000,000,000,000,000,000,000,000,000,000,000,000,000, 000,000,000,000,000,000,000,000,000,000,000,000,000, 000,000,000,000,000,000,000,000,000,000,000,000,000, 000,000,000,000,000,000,000,000,000,000,000,000,000, 000,000,000,000,000,000,000,000,000,000,000,000,000, 000,000,000,000,000,000 universes. Our universe would just happen to be the one with a value of lambda suitable for the existence of creatures that can contemplate their hyper-Copernican existence.

This scenario was, in a way, a logical extension of the argument that Vera Rubin had tried to make with her master's thesis. The Earth is rotating, the solar system is rotating, the galaxy is rotating. Why not the universe? Similarly, the Earth turned out to be just one more planet orbiting the Sun, the Sun one more star in an island universe, the island universe one more galaxy in the universe. And the universe?

By 1973, scientists had named this idea the anthropic principle, and by and large they hated it. Called it "the 'A' word." Wouldn't even discuss it. If any scientific speculation, no matter how wild, ultimately must come down to a prediction, then what prediction could the anthropic principle make? How could you falsify the claim that an unfathomable number of universes exist outside our own? If you can't, critics said, then you had to file the idea under metaphysics—overlooking, perhaps, that it was the metaphysics of midcentury cosmology that had gotten them to a lambda-CDM-plus-inflation universe.

The ongoing resistance to the anthropic principle following the discovery of lambda was similar to an earlier era's discomfort with a homogeneous and isotropic universe following the discovery of the CMB. But then homogeneity and isotropy got an explanation: inflation. A universe that underwent a brief period of extraordinary growth would indeed appear the same wherever you were and wherever you looked.

The anthropic principle was similarly ad hoc—and so what? "Have you heard a better idea?" So wrote no less an authority on initially resisting a homogeneous and isotropic universe than Jim Peebles, in 2003. "I hear complaints that this anthropic principle has been introduced *ad hoc*, to save the phenomenon. But the same is true of $\Lambda$. The cosmological constant is now seen to save quite a few phenomena." As for another explanation for a low value of lambda: "Something may turn up." But it didn't have to; inflation itself had elevated the homogeneity and isotropy of the universe out of the ad hoc and into the inevitable. Maybe inflation did the same for the existence of $10^{500}$ other universes.

Which didn't mean what critics of the anthropic principle often accused it of meaning: the end of physics. While inflation might predict a menagerie of universes, it didn't explain the mechanism that would allow lambda to vary, universe to universe. Theorists would still have to try to work out the physics for that understanding of existence.

And *that* would be their legacy—the legacy of Brian Schmidt, Saul Perlmutter, Adam Riess, and the dozens of other discoverers of evidence for the acceleration of the universe. It wouldn't be personal acrimony, and it wouldn't be changes to their profession's sociology. It would be the revolution in thought that dark energy mandated. Almost certainly this revolution would require the long-awaited union of general relativity and quantum theory. It might involve modifying Einstein's equations. It could feature parallel, intersecting, or a virtually infinite ensemble of universes.

But whatever this revolution wound up being or doing, it would need what speaker after speaker at conference after conference acknowledged, adopting the same shrugging grace and gratitude as the Dicke birds when they learned they'd been scooped, as Vera Rubin when she realized that astronomy had been overlooking most of what was out there: a "new physics."

What greater legacy could a scientist leave a universe?

# Epilogue

THE TRUMPET fanfare began, and then the procession. Up the centre aisle in a Cambridge University combination room — what's called a common room everywhere else — the leaders of the two supernova teams that discovered evidence for dark energy marched in a line of dignitaries. Across a courtyard were the rooms Newton had occupied as a student. Nearby was the observatory where Eddington had plotted the eclipse expedition that validated Einstein's general relativity. At many of the scientific conferences the setting didn't matter, but it did on this occasion: the conferring of the 2007 Gruber Prize in Cosmology. Ten years after noticing something strange in the supernova data, Saul Perlmutter and Brian Schmidt, as well as the entirety of the High-z and SCP collaborations, were beginning to go down in history.

They already had posterity. Whenever the discovery of evidence for cosmic acceleration appeared in a peer-reviewed journal, it would forevermore be accompanied by two citations: Riess, A. G., et al. 1998, *AJ*, 116, 1009; Perlmutter, S., et al. 1999, *ApJ*, 517, 565. But for these recipients of the Gruber Prize in Cosmology, the award ceremony at Cambridge wasn't only about posterity. It was about history, and history was something else. History was posterity in motion.

Schmidt had reached a compromise with the Gruber Foundation regarding the recognition of Adam Riess: The honor would go to everyone — all fifty-one members of the High-z and SCP collaborations. The two teams would split the £300,000 prize; Schmidt and

Perlmutter would each get half of his team's £150,000, and the remaining £75,000 would be divided among the team members. After taxes, the individual awards would amount to maybe £1,000 each, but thirty-five members paid their way to the ceremony in Cambridge. It was probably the first time that so many of them had been in one place, and it might be the last. Perhaps fittingly for a commemoration of a universe that was mostly missing, even an absence was present: Schmidt and Perlmutter included in their joint lecture a PowerPoint slide that recognized the often-overlooked contribution of the Chilean supernova search in the early 1990s—including the name and photograph of José Maza, the mentor to Mario Hamuy who in 1995 withdrew from the programme.

"Our teams, certainly in the U.S., were known for sort of squabbling a bit," Schmidt had said at a press conference in London the day before the awards ceremony. "The accelerating universe was the first thing that our teams ever agreed on," he added, and Perlmutter, standing beside him, laughed. The two of them had extensively discussed in advance how to present a united front, and they had collaborated closely on the choreography of the weekend. For the lecture they gave the day after the ceremony, they worked out a routine; they took turns narrating the history of modern cosmology, sometimes finishing each other's sentences.

*The history of modern cosmology*. A history of something that would have been philosophically laughable to Jim Peebles in 1964, or professionally risky to Michael Turner in 1978, or physically dubious to any number of scientists pre-COBE. Perlmutter and Schmidt were themselves as young as the universe—the one that popped into existence over the course of a phone call in 1965, when theory met observation and a phenomenon in the heavens matched maths on paper. Yet already that cosmology had become a commonplace.

When the anniversary year arrived, the celebrations, not surprisingly, often reflected on the past. "Let's just pause for a second and think how amazingly lucky we are, the stage cosmology has reached," John Peacock, from the University of Edinburgh, said at the spring 2008 Space Telescope Science Institute symposium, "A Decade of Dark Energy." "A poor soldier who died in the trenches in 1914 knew as much about the universe as a caveman." That infantryman lived in a cosmos that was as vast as the stars, but no vaster, and stood still. In the past century, however, our knowledge had grown from one island universe to hundreds of billions of galaxies, from eternally repetitive motions in space to structural evolution over time. And now we even had one more *more:* darkness.

For this reason, the celebrations also often looked not only at how far we'd come but at how far we had to go. "It's not often," STScI director Matt Mountain said at the same meeting, "that astrophysics challenges modern or fundamental physics. Perhaps in the last four hundred years you can count maybe on one hand, perhaps on two, when these instances have occurred. Well, the discovery of the acceleration of the universe a decade ago has handed this generation just one of those opportunities."

Since the invention of the telescope four centuries earlier, astronomers had been able to figure out the workings of the universe simply by observing the heavens and applying some maths, and vice versa. Take the discovery of moons, planets, stars, and galaxies, apply Newton's laws, and you have a universe that runs like clockwork. Take Einstein's modifications of Newton, apply the discovery of cosmic expansion, and you get the Big Bang universe — what Saul Perlmutter once called "a ridiculously simple, intentionally cartoonish picture."

He was sitting in George Smoot's office on the Berkeley campus. Three days earlier Smoot had learned that he had won the 2006 Nobel Prize in Physics for his work on COBE. Bearded, booming, eyes wheeling from adrenaline and lack of sleep, Smoot leaned back in his chair. Perlmutter leaned forward in his.

"Time and time again," Smoot shouted, "the universe has turned out to be really simple."

Perlmutter nodded eagerly. "It's like, why are we able to understand the universe at our level?"

"Exactly. It's a universe for beginners! *The Universe for Dummies!* We're just incredibly lucky that that first try has matched so well."

Would our luck hold? Scientists liked to say that what physics needed was "the next Einstein." But if we took seriously the once-a-millennium quality of the dark-universe revolution — and we had every reason to think we should — then the analogy was inexact. Einstein was our Copernicus, finding the equations that might or might not represent the real—or "real"—universe. The discoverers of dark matter and dark energy were our Galileo, making the observations that validated this universe, though it turned out to be far more elaborately mysterious than we had ever imagined. What science needed now wasn't the next Einstein but the next Newton—someone (or someones, or some collaboration, or some generations-long cathedral of a theory) to codify the maths of this new universe. To unite the physics of the very big with the physics of the very small, just as Newton had united the physics of the celestial with the physics of the terrestrial. To take the observations and make sense of our universe all over again in ways that we couldn't begin to imagine, but that would define our physics and philosophy—our civilization—for centuries to come.

It was this prospect that led cosmologists to regard these originally disturbing discoveries of a universe beyond our senses with fascination and optimism, to view a seeming human limitation as a source of intellectual liberation. "The really hard problems are great," Mike Turner said, "because we know they'll require a crazy new idea." Or as an astronomer told his colleagues at the McMaster conference on dark energy, "If you put the timeline of the history of science before me and I could choose any time and field, this is where I'd want to be."

So: Let there be dark. Let there be doubt, even amidst the certainty. *Especially* amidst the certainties—the pieces of evidence that in one generation transformed cosmology from metaphysics to physics, from speculation to science.

In early 2010, the WMAP seven-year results arrived bearing the latest refinements of the numbers that define our universe. It was 13.75 billion years old. Its Hubble constant was 70.4, and its equation of state ($w$) -0.98, or, within the margin of error, -1.0. And it was flat, consisting of 72.8 percent dark energy, 22.7 percent dark matter, and 4.56 percent baryonic matter (the stuff of us)—an exquisitely precise accounting of the depth of our ignorance. How the story would end remained a mystery, for now and possibly forever. The astronomers who set out to write the final chapter in the history of the universe had to content themselves instead with a more modest conclusion: *To be continued.*

"In a very real sense," Vera Rubin once wrote, "astronomy begins anew." In 1992, the Department of Terrestrial Magnetism moved her to an office in a new building. The photograph of Andromeda came with her, and they promised to put it on her ceiling, but nobody ever got around to it. She didn't care: *The world changes.* Besides, the other office had a low ceiling. There, M31 seemed close enough to touch. In her new office, it would have been out of reach. "The joy and fun of understanding the universe," she continued in that essay, "we bequeath to our grandchildren — and to their grandchildren."

"I have this three-year-old daughter at home," Perlmutter said now, sitting in Smoot's office, "and we're just at that stage where she's asking us, 'Why?' It's pretty obvious that she knows it's a bit of a game. She knows that whatever we say, she can then say, 'Yes, but — why?'" He laughed. "I have the impression that most people don't realize that what got physicists into physics usually is not the desire to understand what we already know but the desire to catch the universe in the act of doing really bizarre things. We *love* the fact that our ordinary intuitions about the world can be fooled, and that the world can just act strangely, and you can just go out and make it good over and over again. 'Do that again! Do that again!'"

Smoot agreed. "They're always testing the limits. And that's what we're doing. We're babies in the universe, and we're testing what the limits are."

If our luck did hold, and another Newton did come along, and the universe turned out once again to be simple in ways we couldn't have previously imagined, then Saul Perlmutter's daughter or Vera Rubin's grandchildren's grandchildren would not be seeing the same sky that they did, because they would not be thinking of it in the same way. They would see the same stars, and they would marvel at the hundreds of billions of galaxies other than our own. But they would sense the dark, too. And to them that darkness would represent a path toward knowledge — toward the kinds of discoveries that we all once called, with understandable innocence, the light.

# Notes

In the notes that follow, interviews are cited by the last name of the subject of the interview in boldface—for example, "**Jones**." The author is grateful to the following for their generosity with their time and knowledge, and apologizes for any omissions.

Eric Adelberger
Daniel Akerib
Greg Aldering
Elena Aprile
Steve Asztalos
Jonathan Bagger
Bradford Benson
Blas Cabrera
Robert Cahn
John Carlstrom
Sean Carroll
Douglas Clowe
Juan Collar
Jodi Cooley
Tom Crawford
Robert P. Crease
Abigail Crites
Richard Ellis
Alex Filippenko
Ann K. Finkbeiner*
Brenna Flaugher
W. Kent Ford
Josh Frieman
Peter Garnavich
Neil Gehrels
Elizabeth George
James Glanz
Gerson Goldhaber
Ariel Goobar
Don Groom
Alan Guth
Mario Hamuy
Gary Hill
Steve Holland

* Some of the citations to Finkbeiner refer to unpublished information from her notes and interviews. The author gives special thanks for this extraordinary courtesy.

William L. Holzapfel
Isobel Hook
Michael T. Hotz
Wayne Hu
Per Olof Hulth
Alex Kim
Darin Shawn Kinion
Robert P. Kirshner
Stuart Klein
Edward W. Kolb
Mark Krasberg
Andrey Kravtsov
Robin Lafever
Bruno Leibundgut
Michael Levi
Eric Linder
Mario Livio
Robert Lupton
Rupak Mahapatra
Stephen P. Maran
Stacy McGaugh
Jeff McMahon
Russet McMillan
Mordehai Milgrom
Richard Muller
Robert Naeye
Heidi Newberg
Peter Nugent
Jeremiah Ostriker
Nikhil Padmanabhan
Robert J. Paulos
P. James E. Peebles
Carl Pennypacker
Saul Perlmutter
Mark Phillips
Paul Preuss
Oriol Pujolas
Adam Riess
Natalie Roe
Leslie Rosenberg
Rob Roser
Vera Rubin
Bernard Sadoulet
Allan Sandage
Kathryn Schaffer
David Schlegel
Brian Schmidt
Lee Smolin
George Smoot
Steph Snedden
Helmuth Spieler
Nicholas Suntzeff
Karl van Bibber
Keith Vanderlinde
Rick van Kooten
Christof Wetterich
Hongsheng Zhao

The following oral histories are cited by the last name of the interview subject and either "AIP" (American Institute of Physics) or "NLA" (National Library of Australia) — for example, "Jones AIP."

Interview of Richard Ellis by Ursula Pavlish, Niels Bohr Library & Archives, American Institute of Physics, College Park, MD, July 27, 2007, AIP.

Interview of Sandra M. Faber by Patrick McCray, July 31, 2002, AIP, http://www.aip.org/history/ohilist/25489.html.
Interview of Alex Kim by Ursula Pavlish, July 31, 2007, AIP.
Interview of Robert Kirshner by Ursula Pavlish, August 3, 2007, AIP.
Interview of Eric Linder by Ursula Pavlish, August 1, 2007, AIP.
Interview of P. James E. Peebles by Christopher Smeenk on April 4 and 5, 2002, http://www.aip.org/history/ohilist/25507.html.
Interview of Vera Cooper Rubin by David DeVorkin, May 9, 1996, AIP, http://www.aip.org/history/ohilist/5920_2.html.
Interview of Brian Schmidt by Ragbir Bhathal, June 15, 2006, Australian Astronomers oral history project, NLA.

page EPIGRAPH

vii "'I know,' said Nick": Hemingway, p. 92.

PROLOGUE

xiii The time had come: **Cooley, Mahapatra.**
xv "We're just a bit": Lawrence Krauss, "New Views of the Universe: Extra Dimensions, Dark Energy, and Cosmic Adventures," Harris Theater, Chicago, Dec. 12, 2005.

1. LET THERE BE LIGHT

3 Dicke helped run: Happer et al., p. 3.
he sometimes got: **Peebles.**
"Well, boys": Wilkinson, p. 204.
deflation or disappointment: **Peebles.**
4 If the caller: Peebles 2009, p. 191.
After the five: Penzias 2009, p. 151.
in 1960: Wilson 1992, p. 463.
Telstar: Penzias 2009, p. 146.
from the fringes: Penzias 2009, p. 147.
Penzias had built: Wilson 2009, p. 164.
New York City: Wilson 1992, p. 475.
The phases: Bernstein, p. 215.
5 They put tape: Wilson 1992, p. 475.
the throat of the horn: Wilson 1992, p. 475.
They caught: Weinberg, p. 47.
five scientists repaired: Penzias 2009, p. 151.
He and the two: Penzias 2009, p. 151.

6 "a beginning": O'Connor, p. 93.
7 "bleeding-heart liberal": Peebles AIP.
On one occasion: Peebles AIP.
When Plato: Crowe, pp. 21–25.
8 Ptolemy: Crowe, pp. 42–49.
13 "The unrestricted": Bondi and Gold, p. 252.
"causes unknown": Hoyle, p. 372.
14 the top student: Peebles AIP.
then one day: Peebles 2009, p. 185.
Friday evenings: Peebles AIP.
in the attic: Lightman and Brawer, p. 218.
"Dicke birds": Happer et al., p. 7.
15 "tiny observable": Pais, p. 273.
16 seemingly homeless: **Finkbeiner**.
17 particularly muggy: Lightman and Brawer, p. 218.
"Why don't you look": **Peebles**.
"Why don't you go": Lightman and Brawer, p. 219.
As a graduate student: Lightman and Brawer, p. 217.
previous years' exams: **Peebles**.
So he studied: Peebles 2009, p. 185.
they presented cosmology: **Peebles**.
General relativity itself: Lightman and Brawer, p. 216.
What appalled him: Peebles 2009, p. 185.
published a paper: Einstein 1917.
18 "They just made": Peebles 2009, p. 186.
"primarily in order": Tolman, p. 332.
*Calculate the acceleration:* Peebles 2009, p. 185.
"Boy, this is silly": Lightman and Brawer, p. 217.
"Having its roots": Kragh, p. 348.
20 A radio antenna: Weinberg, pp. 45–48.
"No problem": **Peebles.**
Peebles delivered: Peebles 2009, p. 191.
former Dicke bird: Peebles 2009, p. 191.
at the Carnegie: K. Turner, p. 184.
mentioned the colloquium: Burke, p. 180.
to an American: Penzias 2009, p. 150.
After a brief: Burke, p. 181.
21 On the flight: Burke, p. 180.
"We have something": Burke, p. 181.
a reciprocal meeting: Penzias 2009, p. 151.

each write a paper: Dicke et al.; Penzias and Wilson.
22 "might outlive": Wilson 1992, p. 476.
"Signals Imply": Walter Sullivan, "Signals Imply a 'Big Bang' Universe," *New York Times,* May 21, 1965.
been in contact: Penzias 2009, p. 152.
"bubbling with excitement": **Peebles**.
A subsequent search: Penzias 1992, pp. 454–56.
In 1948: Gamow 1948, p. 680.
"the temperature": Alpher and Herman, p. 775.
to the cold load: Bernstein, p. 217.
a 1961 article: Ohm, p. 1045.
but that reading: Wilson 2009, p. 169.
23 published a paper: Doroshkevich and Novikov, p. 111.
pointing out that: Penzias 1992, pp. 454–55.
a high metabolism: **Peebles.**
expert downhill skier: Overbye 1992, p. 139.
"a random walk": Ann K. Finkbeiner, "Once Upon the Start of Time," *The Sciences,* Sept./Oct. 1992.
the scholarly process: Overbye 1992, p. 131.
hadn't done his homework: Peebles 2009, p. 192.
repeatedly bounced back: Peebles 2009, p. 191.
June 1965: Peebles 2009, pp. 191–92.
Gamow sent: Penzias 2009, p. 154.
24 "Which is an amazing": Lightman and Brawer, p. 218.

## 2. WHAT'S OUT THERE

25 her father had helped: Lightman and Brawer, pp. 286–87.
linoleum tube: Rubin AIP.
Her second-floor: Rubin 1997, p. 203.
the stars appeared: Rubin AIP.
She memorized: Lightman and Brawer, p. 286.
in high school: Rubin AIP.
At a certain point: Lightman and Brawer, p. 286.
26 had shown her: Rubin AIP.
"Rotating Universe?": Gamow 1946.
Then she heard: Rubin AIP.
data on the 108: Rubin 1951, p. 47.
Then she separated: Rubin AIP.
When an admissions: Irion, p. 960.
27 When a Cornell professor: Rubin AIP.

her paper: Rubin 1951.
it was a master's thesis: Rubin AIP.
"Young Mother": "Young Mother Figures Center of Creation by Star Motions," *Washington Post,* Dec. 31, 1950.
such a novice: Rubin AIP.
"at first sight fantastic": Gamow 1946.

29 She didn't even think: Rubin AIP.
She was standing: **Rubin.**
Her husband shared: Lightman and Brawer, p. 289.

30 No wives: Rubin 1997, p. 187.
taken the job: Rubin AIP.
Rubin had visited: **Rubin.**
Every time she visited: **Rubin.**

31 embarrassed her: Rubin 1997, p. 189.
his own lecture: Gingerich.
two kinds of geniuses: Lightman and Brawer, p. 294.
"Is there a": Rubin AIP.
"On a large": Rubin 1997, p. 198.

32 she had received: Rubin 1997, p. 199.
"From an analysis": de Vaucouleurs, p. 30.
a thick German: Rubin AIP.

33 her thesis: Rubin 1954.
went on a tour: Rubin 1997, p. 87.
"Galaxies may be": Rubin AIP.
motions of 888: Rubin 1997, p. 116.
[Footnote] When the editor: **Rubin.**
[Footnote] the resulting paper: Rubin et al. 1962.
galactic anticenter: Rubin 1997, p. 157.
[Footnote] "This," the astronomer: Rubin 1997, p. 156.

34 visit her friend: **Rubin**.
"He couldn't have": **Rubin**.
She said she could: **Rubin**.
had the choice: **Rubin**.
she noticed that: **Rubin**.

35 Instead, it converted: Bartusiak, p. 208.
the instrument reduced: Rubin 2006, p. 8.
On the whole: Lightman and Brawer, p. 295.

36 She needed to find: Lightman and Brawer, pp. 295–96.
"Within a galaxy": Rubin 1997, p. 1.
rather than pushing: Rubin 1997, p. 89.

37 he needed: Rubin 1997, pp. 131–32.
39 a graduate student: Faber AIP.

### 3. CHOOSING HALOS

42 [Footnote] The name: Peebles 1969, p. 18.
CDC 3600: Peebles 1970, p. 15.
many magnitudes: Overbye 1992, p. 142.
security detail: **Finkbeiner**.
300 points: Peebles 1970, p. 13.
43 felt that the field: Peebles 1999, p. 1067.
44 "Gravitation and Space Science": Dicke and Peebles.
early March 1965: Dicke and Peebles, p. 419.
a note in proof: Dicke and Peebles, p. 460.
Dicke handled: **Peebles**.
in the introduction: Dicke and Peebles, p. 419.
In the first paragraph: Dicke and Peebles, p. 442.
"The moral of this section": Dicke and Peebles, p. 454.
this paper ran: Peebles 1965.
45 During one presentation: Boynton, p. 303.
"The radiation": Peebles 1969, p. 20.
"distributions of mass": Peebles 1965, p. 1317.
In the fall semester: **Peebles**.
46 "The great goal": Peebles 1974, p. vii.
"You measure": Peebles AIP.
a "candidate": Peebles, 1974, p. x.
"the very broad": Peebles 1974, p. xi.
47 Jeremiah Ostriker: **Ostriker**.
48 For the first simulation: **Peebles.**
In 1933: Zwicky 1933.
"a great mass": Smith, p. 23.
"The discrepancy seems": Rubin 2003, p. 2.
49 Peebles himself had regarded: **Peebles.**
"the halo masses": Ostriker and Peebles, p. 467.
50 "There are reasons": Ostriker et al., p. L1.
Just brilliant: **Rubin**.
Peebles hardly noticed: **Peebles**.
such intense hostility: Lightman and Brawer, p. 275.
51 "well discussed": Rubin et al. 1973, p. L111.
"The results": Rubin et al. 1973, p. L111.
two papers: Rubin et al. 1976a, 1976b.

She didn't like: **Rubin**.
She would say: Lightman and Brawer, p. 305.
above the entrance: **Rubin**.
Besides, she and Ford: **Ford**.
52 In 1978 Ford: Rubin et al. 1978.
In 1975 Roberts: Rubin 2003, p. 5.
A 1978 survey: Bosma.
that radio observations: **Peebles**.
that optical data: Lightman and Brawer, p. 303.
53 "Is there more": Faber and Gallagher, p. 135.
"After reviewing": Faber and Gallagher, p. 182.
"many astronomers": Rubin 2003, p. 3.
left most astronomers: Rubin 2003, p. 3.
"Nobody ever": Overbye 1992, p. 307.
"recognizing that": Rubin 1983, p. 1344.

4. GETTING IN THE GAME

58 first director of the center: **Sadoulet**.
59 "that all the particles": http://www.gutenberg.org/files/10350/10350-8.txt.
60 "And so that": Newton, p. 940.
"the small velocities": Einstein 1917, p. 26.
"at present unknown": Einstein 1917, p. 24.
63 "probably a number": Baade, p. 287.
Lawrence Berkeley National Laboratory: http://www.lbl.gov/Science-Articles/Research-Review/Magazine/1981/.
64 "We are an": **Muller.**
One day Alvarez: **Muller**.
had planted: Colgate et al., pp. 565–66.
65 University of Wisconsin: Colgate et al., p. 572.
FORTRAN: Colgate et al., p. 572.
Alvarez looked up: **Muller**.
66 Year after year: **Muller**.
Science courses: **Perlmutter**.
68 pair of paleontologists: Raup and Sepkoski.
The following year: Davis et al. 1984.
his thesis: Perlmutter 1986.
the two projects: **Perlmutter**.
69 In 1981: Kare et al.
Muller himself: **Muller**.

Pennypacker commissioned: **Newberg.**
70 "they are rare": Perlmutter et al. 1995b, p. 2.
already testing: **Perlmutter.**
From 1986 to 1988: Kirshner, pp. 168–70.
71 For the members: **Perlmutter.**
The contractor: Newberg, p. 23.
without a filter: **Newberg.**
72 take the computer: Newberg, p. 18.
Before long: **Perlmutter.**
73 Every few months: **Perlmutter.**
"Look, two": **Muller.**
Newcomers: **Goobar.**
One graduate student: Kim AIP.
A postdoc: **Goobar.**
And then Pennypacker: **Pennypacker.**
74 Muller thought: **Muller.**
Perlmutter stayed: **Perlmutter.**
75 began to think: **Pennypacker.**
76 Ellis snapped: **Perlmutter.**
On August 29, 1992: **Ellis.**

## 5. STAYING IN THE GAME

77 Nicholas Suntzeff: **Suntzeff.**
Schmidt mentioned: **Schmidt.**
78 in Suntzeff's experience: **Suntzeff.**
"How long": **Schmidt.**
Schmidt disappeared: **Garnavich.**
79 privately review: **Ellis.**
publicly write: Kirshner, pp. 170–71.
serve as referee: Kirshner, p. 185.
to know photometry: Kirshner, p. 185.
80 In the 1980s: Kirshner, pp. 180–81.
In 1989, Muller: Kirshner, pp. 178–80.
81 "You must understand": Sandage 1987, p. 3.
"Essentially": Walter Sullivan, "A Yardstick for the Universe?" *New York Times,* Oct. 9, 1984.
at least two classes: Kirshner, pp. 37–38, 160–62.
82 three supernovae: **Kirshner.**
83 including Kirshner: Uomoto and Kirshner.
"There is still": Newberg, p. 13.

84 a "realist": Kirshner, p. 167.
In his role: Kirshner, p. 183.
"They hadn't": Kirshner, p. 190.
In high school: **Suntzeff.**
85 "There are only": **Suntzeff.**
influential article: Sandage 1970.
86 as a "preliminary": Hubble 1936, p. 20.
"plates of Moses": Overbye 1992, p. 27.
87 Suntzeff and Sandage: **Suntzeff.**
He had lost: **Sandage.**
89 Suntzeff was already: **Suntzeff, Phillips.**
91 Mario Hamuy: **Hamuy.**
Bruno Leibundgut: **Leibundgut.**
93 The correlation: **Phillips.**
95 He wrote some: **Schmidt.**
96 The first night: **Leibundgut.**
16-pixel-by-16-pixel: **Schmidt.**
97 Late that night: **Leibundgut.**

6. THE GAME

98 The Berkeley team: **Perlmutter.**
99 "demonstration runs": Perlmutter et al. 1995b, p. 4.
"pilot search": Newberg, p. 105.
seemed oblivious: Kirshner, p. 183.
"No!": Joel Primack, "UCLA Eighth International Symposium on Sources and Detection of Dark Matter and Dark Energy in the Universe," Marina del Rey, Feb. 20–22, 2008.
had demonstrated: **Phillips.**
100 The team often: **Perlmutter.**
They called it: Perlmutter et al. 1995, p. 4.
101 "'I just heard'": **Kirshner.**
"working together": **Kirshner.**
"dimness": Riess et al. 1996, p. 90.
102 "I've spent": **Leibundgut.**
"Crap!": **Suntzeff.**
103 "on the smell": **Schmidt.**
"anarchy": **Suntzeff.**
104 1994 proposal: **Riess.**
a reminder: **Leibundgut.**
The paper: Leibundgut et al.

"We can only": **Suntzeff.**
105 *No big guns:* **Suntzeff.**
Mario Hamuy: **Hamuy.**
106 Hamuy's: Hamuy et al., p. 1.
Riess's: Riess et al. 1995, p. L17.
"How could I": **Hamuy.**
107 Everyone in astronomy: Overbye 1992, p. 188.
Even Sandage: **Suntzeff.**
"The answer": Overbye 1992, p. 278.
He had received: **Suntzeff.**
108 Hamuy himself: **Hamuy.**
"We have to": **Phillips.**
"as if blood": **Suntzeff.**
*that* paper: Riess et al. 1995b.
"I'm a staff astronomer": **Suntzeff.**
109 "Yeah": **Suntzeff.**
His argument: **Kirshner.**
And not only: **Schmidt.**
110 apply for time: **Leibundgut.**
To make the situation: **Suntzeff.**
111 In January 1996: **Perlmutter.**
112 The high resolution: Kirshner, p. 203.
A month later: **Perlmutter.**
proposal came up: **Kirshner.**
He had served: **Perlmutter.**
began to object: **Kirshner.**
into his office: **Kirshner, Phillips, Suntzeff.**
113 Maybe they all: **Suntzeff.**
Bob Cahn: **Cahn.**
115 help justify: **Garnavich.**
announced the results: Perlmutter et al. 1997.
Gerson Goldhaber: **Goldhaber.**
Adam Riess: **Riess.**

## 7. THE FLAT UNIVERSE SOCIETY

119 On Monday evenings: Rex Graham, "Deep-Dish Cosmologists," *Astronomy,* June 2001.
DuPage County: **Kolb, Turner.**
The classroom: Graham, "Deep-Dish Cosmologists."
"backup" hamburgers: **Turner.**

The topics: Graham, "Deep-Dish Cosmologists."

121 classes sometimes: **Turner.**

"less than elegant": **Kolb.**

Cheech and Chong: **Turner.**

Oreos-and-beer: Anton, p. 103.

"I don't believe": **Perlmutter.**

"butchers its young": David H. Freeman, "Particle Hunters," *Discover*, Dec. 1991.

*Don't be afraid:* **Turner.**

122 soon dropped out: Steve Nadis, "The Lost Years of Michael Turner," *Astronomy,* Apr. 2004.

Turner audited: **Turner.**

123 "Why don't you": Nadis, "Lost Years of Michael Turner."

"That early-universe": **Turner.**

this lesson: **Turner.**

His colleagues: David Overbye, "Remembering David Schramm, Gentle Giant of Cosmology," *New York Times,* Feb. 10, 1998.

124 but the combination: **Turner.**

In October 1981: Guth 1998, p. 223.

"< 1 sec.": Guth 1998, p. 223.

He figured: **Turner.**

125 "assumes certain": Guth 1998, p. 223.

Late in the evening: Guth 1998, pp. 167–87.

126 The suggestion followed: Guth 1998, pp. 12–14.

"Is the Universe": Tryon, pp. 396–97.

127 "the universe is": Guth 1998, p. 15.

Guth realized: Guth 1998, pp. 167–87.

128 At three minutes: Smoot and Davidson, p. 161.

129 "Yeah," Guth thought: **Guth.**

Guth's paper: Guth 1981, pp. 347–56.

Seventeen of the: Wilczek 1985, p. 475.

130 Guth's original idea: Guth 1998, pp. 202–10.

In 1973 Hawking: Overbye 1992, pp. 111–15.

131 "traveling circus": Guth 1998, p. 211.

In early 1982: Guth 1998, pp. 211, 215–18.

"informal discussion": Guth 1998, p. 223.

day trips: Guth 1998, p. 232.

croquet and tea: **Turner.**

"a workshop where": **Turner.**

During his talk: Guth 1998, p. 231.

"Child's play": **Turner**.
132 the summer of 1981: K. Turner, p. 12.
133 tour de force: Burbidge et al.
134 logo that showed: Kolb et al., back cover.
T-shirts: Kolb et al., p. 2.
organized a jog: Overbye 1992, p. 214.
"Buffalo Class": Kolb et al., p. 626.
"Cosmology in the": Kolb et al., p. 622.
"Whatever future": Kolb et al., p. 625.
"Perhaps future": Kolb and Turner, p. 498.
"Despite being": Kolb and Turner, p. xix.
135 against a bar: Graham, "Deep-Dish Cosmologists."
In 1976: Smoot and Davidson, pp. 117–43.
136 "the superclustering phenomenon": Davis, p. 111.
survey of galaxies: Geller and Huchra.
137 "frothy": Davis et al. 1982, p. 423.
John Mather: Mather and Boslough, p. 225.
Or not: Glanz 1995.
138 He wanted to know: **Turner.**
139 Swearingen SW-3: Eric Pace, "David Schramm, 52, Expert on the Big Bang," *New York Times*, Dec. 22, 1997.
Turner said: **Turner**.

8. HELLO, LAMBDA

140 Perlmutter had flown: Anton, p. 115.
seatback phone: **Perlmutter**.
in a paper: Perlmutter et al. 1998.
141 "For the first time": Kathy Sawyer, "Universe Will Keep Expanding Forever, Research Teams Say," *Washington Post,* Jan. 9, 1998.
at a table: **Perlmutter**.
for an hour: Anton, p. 116.
Michael Turner: **Turner.**
the panels: http://www.supernova.lbl.gov/.
142 "Dave would have": Dennis Overbye, "Remembering David Schramm, Gentle Giant of Cosmology," *New York Times,* Feb. 10, 1998.
143 "hypothetical and": Bondi and Gold, p. 263.
144 published a paper: Petrossian et al.
"the most plausible": Gunn and Tinsley.
agree that a model: **Turner**.

Among them was: Wilczek 1985, p. 479.
145 Wilczek ended: Wilczek 1985, p. 480.
work on a paper: M. Turner et al. 1984.
Turner liked to say: **Turner.**
"heart of hearts": Finkbeiner, p. 320.
prided himself: Ann K. Finkbeiner, "Once Upon the Start of Time," *The Sciences,* Sept./Oct. 1992, p. 10.
146 "What's best": Finkbeiner, "Once Upon the Start of Time," p. 8.
had emerged: **Peebles.**
a 1983 paper: Davis and Peebles.
"High mass": **Peebles.**
Their conclusion: Davis and Peebles.
"we lose the": Peebles 1984, p. 444.
"It's ugly": Finkbeiner, p. 319; **Turner.**
"It's an addition": Finkbeiner, p. 319.
"Considering the observations": Finkbeiner, p. 320.
The paper met: **Peebles.**
147 "WHY A COSMOLOGICAL": Carroll et al., p. 501.
"The Observational Case": Ostriker and Steinhardt.
Turner again: Krauss and Turner.
with a joke: Glanz 1996, p. 1168.
On one side: Kirshner 2002, p. 193.
148 "(for $\Lambda = 0$)": Kim et al., p. 4.
"(for $\Lambda = 0$)": Perlmutter et al. 1995a, p. L42.
"If we assume": Goldhaber et al., p. 7.
subject of a paper: Goobar and Perlmutter.
their assumption: **Goobar.**
that's what Perlmutter: **Perlmutter.**
"This could be": Overbye 1996, p. 1428.
"jugular science": Overbye, "Remembering David Schramm."
"I don't think": **Glanz.**
"I am anxiously": James Glanz, "Accelerating the Cosmos," *Astronomy,* Oct. 1999.
submitted their data: Perlmutter et al. 1997.
149 a can-do: **Perlmutter.**
On June 30: **Nugent.**
151 submitted its paper: Perlmutter et al. 1998.
152 "suggests that matter": Garnavich et al., p. L53.
Gerson Goldhaber: **Goldhaber.**
"Perhaps the most": "SCP Meeting Notes, 1997 October 08."

153 "antagonistic": **Goldhaber**.
In case: **Perlmutter**.
The two of them: **Riess**.

155 "Adam is sloppy": **Schmidt.**
developed a routine: **Riess**.
signed their e-mails: Michael Anft, "Chasing the Great Beyond," *Johns Hopkins Magazine*, February 2008.
you're Stephen Hawking: **Riess**.
the phone would ring: **Schmidt**.
was a graduate student: **Riess.**

156 Alex Filippenko: **Filippenko**.
He disliked: **Filippenko**.

157 "Man": **Filippenko.**

158 By January 4: **Riess.**
When Pete Garnavich: **Garnavich.**
That evening: **Riess.**

161 "Perlmutter bowled": Sawyer, "Universe Will Keep Expanding Forever."
had written about: Glanz 1995.
"perhaps boosted": Glanz 1997.
"a quantum-mechanical": Glanz 1998a.

162 Filippenko would be: **Filippenko**.

163 "evidence": Anton, p. 117.

## 9. THE TOOTH FAIRY TWICE

164 He was walking: **Turner**.
in 1976: Dennis Overbye, "Remembering David Schramm, Gentle Giant of Cosmology," *New York Times,* Feb. 10, 1998.

165 broke the story: Glanz 1998b.
"a preponderance": **Glanz**.

166 1,600-word feature: John Noble Wilford, "Wary Astronomers Ponder an Accelerating Universe," *New York Times*, Mar. 3, 1998.
lab press release: "Supernova Cosmology Project Named in *Science* Magazine's Breakthrough of the Year," Dec. 17, 1998, http://www.osti.gov.news/releases98/decpr/pr98192.htm.
"Basically": John Noble Wilford, "In the Light of Dying Stars, Astronomers See Intimations of Cosmic Immortality," *New York Times*, Apr. 21, 1998.
"Observational Evidence": Riess et al. 1998.

A straw poll: The Editors, "Revolution in Cosmology," *Scientific American*, January 1999.

167 "Their highest": **Turner.**

"It made": **Turner.**

168 "Admit it": **Schmidt.**

"Cosmology Solved? Maybe": M. Turner 1998a.

straightforward "Cosmology Solved?": M. Turner 1998b.

169 "distasteful": John Noble Wilford, "New Findings Help Balance the Cosmological Books," *New York Times*, Feb. 9, 1999.

When he thought: Wilford, "New Findings Help Balance."

"half enthusiast": Wilford, "New Findings Help Balance."

170 "ridiculous": **Turner.**

"purposefully provocative": M. Turner 1999.

171 Adam Riess: **Riess.**

Not only did: Carroll et al., pp. 503–4.

"No worker": Carroll et al., p. 504.

172 "you wouldn't be able": **Turner.**

would have cooled: Carroll et al., p. 503.

173 "I have not": Thompson, p. 1065.

"superfluous": Stachel, p. 124.

"You observational": **Filippenko.**

174 "Nobody has ever": **Riess.**

176 made their point: Gilliland et al.

He couldn't stop: **Riess.**

178 "astronomy of the invisible": Dennis Overbye, "From Light to Darkness: Astronomy's New Universe," *New York Times*, Apr. 10, 2001.

179 Among the observers: "The Dark Universe: Matter, Energy, and Gravity," Space Telescope Science Institute, Baltimore, Apr. 2–5, 2001.

as she pointed out: Rubin 2003, p. 1.

## 10. THE CURSE OF THE BAMBINO

185 Karl van Bibber: **van Bibber.**

catching an axion: van Bibber and Rosenberg, p. 31.

186 "cold planets": Rubin 1997, p. 128.

Paczynski suggested: Paczynski 1986a, 1986b.

187 "Of course": Gates, p. 71.

"the probability": Gates, p. 72.

188 discovered that deuterium: Riordan and Schramm, p. 81.
"baryometer": Turner 1999.
By similar reasoning: Riordan and Schramm, pp. 81–83.
189 and perhaps higher: Riordan and Schramm, pp. 85–86.
190 a "Great Wall": Geller and Huchra.
Two-degree-Field: Peacock et al.
Sloan Digital: Abazajian et al.
Galaxies formed first: Finkbeiner 1996.
Cosmic Evolution Survey: Massey et al.
191 closely parsed: **Clowe**.
192 "you could put": **Turner**.
194 a group of twelve: **Mahapatra**.
"Of course": **Cabrera**.
196 "We would have totally": **Cooley**.
"Dark matter discovered?": http://monkeyfilter.com/link.php/17049.
"Has Dark Matter": http://www.popsci.com/science/article/2009-12/evidence-dark-matter-emerges-worlds-most-sensative-detector.
"Rumor has it": http://greenteabreak.com/2009/12/08/rumor-has-it-that-the-first-dark-matter-particle-has-been-found/.
"*¿Se ha descubierto*": http://www.migui.com/ciencias/fisica/¿se-ha-descubierto-la-materia-oscura-en-el-cdms.html.
"*Pátrání po supersymetrické*": http://www.scinet.cz/patrani-po-supersymetricke-skryte-hmote.html.
"みんな大好き(か": http://ameblo.jp/physics/entry-10409525072.html.
"the Thursday speakers": http://resonaances.blogspot.com/2009/12/little-update-on-cdms.html.
197 *Discover* magazine: http://blogs.discovermagazine.com/cosmic-variance/2009/12/17/dark-matter-detected-or-not-live-blogging-the-seminar/.
"The results": http://titus.stanford.edu/public/movies/vmt6ud.mov.
"I'm in love": **Rosenberg**.
198 "I called this particle": Wilczek 1991.
"vanishingly small": **van Bibber**.
In 1989: **van Bibber**.
200 Juan Collar: **Collar**.
201 "sort of a rude": **van Bibber**.

### 11. THE THING

203 William L. Holzapfel: **Holzapfel.**
204 For Holzapfel: **Holzapfel.**
205 acceleration paper: Perlmutter et al. 1999.
206 "But I don't": **Bennett.**
207 The first question: National Research Council, p. 2.
208 "In a Universe": Michael S. Turner, "Dark Energy and the New Cosmology," http://supernova.lbl.gov/~evlinder/turner.pdf.
in January 2002: National Research Council, p. 184.
"the most vexing": National Research Council, p. 144.
210 Dark Universe meeting: Riess et al. 2001.
In 2004: Riess et al. 2004.
and 2006: Riess et al. 2007.
Peebles had noted: Peebles and Yu.
mapped the locations: Eisenstein et al.
211 "I hit 'go'": **George.**
214 required shipping: William Mullen, "Dark Energy in the Spotlight," *Chicago Tribune,* Dec. 31, 2007.
And because much: **Crawford, McMahon.**
215 Twice a day: **Vanderlinde.**
started making noises: **Crawford, McMahon.**
216 signed off on a paper: Staniszewski et al.

### 12. MUST COME DOWN

219 weeklong meeting: "Origins of Dark Energy," Origins Institute, McMaster University, Hamilton, Ontario, May 14–17, 2007; Perimeter Institute, Waterloo, Ontario, May 18–20, 2007.
220 "I'm the last": **Schmidt.**
222 "Experience remains": Einstein 1934, p. 274.
223 "I hold it": Einstein 1934, p. 274.
"Saul thinks": **Suntzeff.**
"Saul is going": **Filippenko.**
224 "big enough": **Riess.**
"mentioned NASA": Mather and Boslough, p. 236.
"appear to be": **Filippenko.**
225 Filippenko's personal: Filippenko 2001.
Kirshner published: Kirshner.
"Well, it tells me": **Cahn.**
"The High-z accounts": **Groom.**
226 "Dear Saul": **Kirshner.**

Over the years: Schmidt NLA.
"We announced": http://www.shawprize.org/en/laureates/2006/astronomy/Perlmutter_Riess_Schmidt/autobiography/Perlmutter.html.
227 "Thoughts on": http://www.cfa.harvard.edu/~rkirshner/whowhatwhen/Thoughts.htm.
"Dear Bob": **Perlmutter.**
Kirshner addressed: **Kirshner.**
228 the feeling among some: **Filippenko.**
"My team": **Goldhaber.**
ran an article: Sharon Begley, "In 'Dark Energy,' Cosmic Humility," *Newsweek,* Oct. 1, 1998.
Part One began: http://newscenter.lbl.gov/news-releases/2007/12/12/dark-energys-10th-anniversary-2/.
"Fundamentalist Physics": White.
229 another cosmology meeting: American Astronomical Society meeting, Minneapolis, May 29–June 2, 2005.
230 "We have *not*": American Astronomical Society meeting, Washington, DC, Jan. 6–10, 2002.
"deep theory": **Riess.**
Perlmutter liked to begin: **Perlmutter.**
a friend: Glazebrook.
yet one more conference: "New Views of the Universe" symposium, Chicago, Dec. 8–13, 2005.
"heyday for talking": **Carroll.**
232 "not worth it!": "A Decade of Dark Energy," Space Telescope Science Institute, Baltimore, May 5–8, 2008.
"standing out": **Schmidt.**
"then it's a number": **Livio.**
one of the dark-energy conferences: "A Decade of Dark Energy," Space Telescope Science Institute, Baltimore, May 5–8, 2008.
234 even its defenders: **McGaugh.**
235 less than a millimeter: **Adelberger.**
237 "Have you heard": Peebles 2003, p. 4.
238 didn't explain the mechanism: John Peacock, "A Decade of Dark Energy."

## EPILOGUE

239 The trumpet fanfare: Video, Peter and Patricia Gruber Foundation, New York.

240 "Let's just pause": "A Decade of Dark Energy," Space Telescope Science Institute.
241 George Smoot's: **Smoot.**
242 "The really hard problems": **Turner.**
"If you put": Andreas Albrecht, "Origins of Dark Energy," McMaster University.
243 "In a very real": Rubin 1997, p. 129.
*The world changes:* **Rubin.**

# Works Cited

Citations of works of central importance to this book list all authors. Citations of works of more tangential interest by more than two authors list only the first author, followed by "et al."

Abazajian, Kevork N., et al. 2009. *Astrophysical Journal Supplement* 182: 543–58.

Alcock, C., et al. 1993. "Possible Gravitational Microlensing of a Star in the Large Magellanic Cloud." *Nature* 365: 621–23.

Alpher, R. A., and R. C. Herman. 1948. "Evolution of the Universe." *Nature* 162: 774–75.

Anton, Ted. 2001. *Bold Science: Seven Scientists Who Are Changing Our World.* New York: W. H. Freeman.

Aubourg, E., et al. 1993. "Evidence for Gravitational Microlensing by Dark Objects in the Galactic Halo." *Nature* 365: 623–25.

Australian Astronomers Oral History Project. Interview of Brian Schmidt by Ragbir Bhathal, June 15, 2006. National Library of Australia.

Baade, W. 1938. "The Absolute Photographic Magnitude of Supernovae." *Astrophysical Journal* 88: 285–304.

Bartusiak, Marcia. 1993. *Through a Universe Darkly: A Cosmic Tale of Ancient Ethers, Dark Matter, and the Fate of the Universe.* New York: HarperCollins.

Bernstein, Jeremy. 1986. *Three Degrees above Zero.* New York: New American Library.

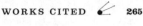

Bernstein, Jeremy, and Gerald Feinberg, eds. 1986. *Cosmological Constants: Papers in Modern Cosmology*. New York: Columbia University Press.

Bondi, H., and T. Gold. 1948. "The Steady-State Theory of the Expanding Universe." *Monthly Notices of the Royal Astronomical Society* 108: 252–70.

Bosma, A. 1978. "The Distribution and Kinematics of Neutral Hydrogen in Spiral Galaxies of Various Morphological Types." PhD diss., Groningen University, Groningen, Neth.

Boynton, Paul. 2009. "Testing the Fireball Hypothesis." In *Finding the Big Bang*, ed. P. James E. Peebles, Lyman A. Page, Jr., and R. Bruce Partridge. New York: Cambridge University Press.

Burbidge, E. Margaret, G. R. Burbidge, William A. Fowler, and F. Hoyle. 1957. "Synthesis of the Elements in Stars." *Reviews of Modern Physics* 29: 547–650.

Burke, Bernard F. 2009. "Radio Astronomy from the First Contacts to the CMBR." In *Finding the Big Bang*, ed. P. James E. Peebles, Lyman A. Page, Jr., and R. Bruce Partridge. New York: Cambridge University Press.

Carroll, Sean M., William H. Press, and Edwin L. Turner. 1992. "The Cosmological Constant." *Annual Review of Astronomy and Astrophysics* 30: 499–542.

Colgate, Stirling A., Elliott P. Moore, and Richard Carlson. 1975. "A Fully Automated Digitally Controlled 30-inch Telescope." *Publications of the Astronomical Society of the Pacific* 87: 565–75.

Crowe, Michael J. 1990. *Theories of the World from Ptolemy to Copernicus*. New York: Dover.

Davis, M. 1982. "Galaxy Clustering and the Missing Mass." *Philosophical Transactions of the Royal Society of London, Series A, Mathematical and Physical Sciences* 307: 111–19.

Davis, M., J. Huchra, D. W. Latham, and J. Tonry. 1982. "A Survey of Galaxy Redshifts, II: The Large Scale Space Distribution." *Astrophysical Journal* 253: 423–45.

Davis, Marc, and P. J. E. Peebles. 1983. "A Survey of Galaxy Redshifts, V: The Two-Point Position and Velocity Correlations." *Astrophysical Journal* 267: 465–82.

Davis, Marc, Piet Hut, and Richard A. Muller. 1984. "Terrestrial Catastrophism: Nemesis or Galaxy?" *Nature* 308: 715–17.

de Vaucouleurs, Gérard. 1953. "Evidence for a Local Supergalaxy." *Astronomical Journal* 58: 30–32.

Dicke, R. H., and P. J. E. Peebles. 1965. "Gravitation and Space Science." *Space Science Reviews* 4: 419–60.

Dicke, R. H., P. J. E. Peebles, P. G. Roll, and D. T. Wilkinson. 1965. "Cosmic Black-Body Radiation." *Astrophysical Journal* 142: 414–19.

Doroshkevich, A. G., and I. D. Novikov. 1964. "Mean Density of Radiation in the Metagalaxy and Certain Problems in Relativistic Cosmology." *Soviet Physics Doklady* 9: 111.

Einstein, Albert. 1917. "Cosmological Considerations on the General Theory of Relativity." Reprinted in *Cosmological Constants: Papers in Modern Cosmology*, ed. Jeremy Bernstein and Gerald Feinberg. New York: Columbia University Press, 1986.

———. 1934. "On the Method of Theoretical Physics." Reprinted in *Ideas and Opinions*, ed. Carl Seelig. New York: Bonanza Books, 1954.

Eisenstein, Daniel, et al. 2005. "Detection of the Baryon Acoustic Peak in the Large-Scale Correlation Function of SDSS Luminous Red Galaxies." *Astrophysical Journal* 633: 560–74.

Faber, S. M., and J. S. Gallagher. 1979. "Masses and Mass-to-Light Ratios of Galaxies." *Annual Review of Astronomy and Astrophysics* 17: 135–87.

Filippenko, Alexei V. 2001. "Einstein's Biggest Blunder? High-Redshift Supernovae and the Accelerating Universe." *Publications of the Astronomical Society of the Pacific* 113: 1441–48.

Finkbeiner, Ann. 1992. "Cosmologists Search the Universe for a Dubious Panacea." *Science* 256: 319–20.

Gamow, G. 1946. "Rotating Universe?" *Nature* 158: 549.

———. 1948. "The Evolution of the Universe." *Nature* 162: 680–82.

Garnavich, Peter M., Robert P. Kirshner, Peter Challis, John Tonry, Ron L. Gilliland, R. Chris Smith, Alejandro Clocchiatti, Alan Diercks, Alexei V. Filippenko, Mario Hamuy, Craig J. Hogan, B. Leibundgut, M. M. Phillips, David Reiss, Adam G. Riess, Brian P. Schmidt, J. Spyromilio, Christopher Stubbs, Nicholas B. Suntzeff, and Lisa Wells. 1998. "Constraints on Cosmological Models from Hubble Space Telescope Observations of High-z Supernovae." *Astrophysical Journal* 493: L53–L57.

Gates, Evalyn. 2009. *Einstein's Telescope: The Hunt for Dark Matter and Dark Energy in the Universe.* New York: W. W. Norton.

Geller, Margaret J., and John P. Huchra. 1989. "Mapping the Universe." *Science* 246: 897–903.

Gibbons, G. W., S. W. Hawking, and S. T. C. Siklos, eds. 1985. *The Very Early Universe.* Cambridge: Cambridge University Press.

Gilliland, Ronald L., Peter E. Nugent, and M. M. Phillips. 1999. "High-Redshift Supernovae in the Hubble Deep Field." *Astrophysical Journal* 521: 30–49.

Gingerich, Owen. 1994. "The Summer of 1953: A Watershed for Astrophysics." *Physics Today* 47: 34–40.

Glanz, James. 1995. "To Learn the Universe's Fate, Observers Clock Its Slowdown." *Science* 269: 756–57.

———. 1996. "Debating the Big Questions." *Science* 273: 1168–70.

———. 1997. "New Light on Fate of the Universe." *Science* 278: 799–800.

———. 1998a. "Exploding Stars Point to a Universal Repulsive Force." *Science* 279: 651–52.

———. 1998b. "Astronomers See a Cosmic Antigravity Force at Work." *Science* 279: 1298–99.

Glazebrook, Karl. 2006. "The WiggleZ Survey." PowerPoint presentation.

Goldhaber, Gerson. 1994. "Discovery of the Most Distant Supernovae and the Quest for Omega." Lawrence Berkeley National Laboratory, LBNL Paper LBL-36361.

Goobar, Ariel, and Saul Perlmutter. 1995. "Feasibility of Measuring the Cosmological Constant Λ and Mass Density Ω Using Type Ia Supernovae." *Astrophysical Journal* 450: 14–18.

Gunn, James E., and Beatrice M. Tinsley. 1975. "An Accelerating Universe." *Nature* 257: 454–57.

Guth, Alan H. 1981. "The Inflationary Universe: A Possible Solution to the Horizon and Flatness Problems." *Physical Review D* 23: 347–56.

———. 1998. *The Inflationary Universe: The Quest for a New Theory of Cosmic Origins.* New York: Basic Books.

Hamuy, Mario, M. M. Phillips, José Maza, Nicholas B. Suntzeff, R. A. Schommer, and R. Aviles. 1995. "A Hubble Diagram of Distant Type Ia Supernovae." *Astronomical Journal* 109: 1–13.

Happer, W., P. J. E. Peebles, and David Wilkinson. 1999. "Robert Henry

Dicke, 1916–1997." *Biographical Memoirs* 77. Washington, DC: National Academy Press.

Hemingway, Ernest. 1966. "Indian Camp." In *The Short Stories of Ernest Hemingway*. New York: Charles Scribner's Sons.

Hewitt, Adelaide, Geoffrey Burbidge, and Li Zhi Fang, eds. 1987. *Observational Cosmology*. Proceedings of the 124th Symposium of the International Astronomical Union, Beijing, China, August 25–30, 1986. Dordrecht, Neth.: D. Reidel.

Hoyle, F. 1948. "A New Model for the Expanding Universe." *Monthly Notices of the Royal Astronomical Society* 108: 372–82.

Hubble, Edwin. 1936. *The Realm of the Nebulae*. New Haven: Yale University Press.

Irion, Robert. 2002. "The Bright Face behind the Dark Sides of Galaxies." *Science* 295: 960–61.

Kare, Jordin T., Carlton R. Pennypacker, Richard A. Muller, Terry S. Mast, Frank S. Crawford, and M. Shane Burns. 1981. "The Berkeley Automated Supernova Search." Presented at the North American Treaty Organization/Advanced Study Institute on Supernovae, Cambridge, England, June 28–July 10, 1981.

Kim, A., et al. 1995. "K Corrections for Type Ia Supernovae and a Test for Spatial Variation of the Hubble Constant." Presented at the NATO Advanced Study Institute Thermonuclear Supernovae Conference, Aiguablava, Spain, June 20–30, 1995.

Kirshner, Robert P. 2002. *The Extravagant Universe: Exploding Stars, Dark Energy, and the Accelerating Cosmos*. Princeton, NJ: Princeton University Press.

Kolb, Edward W., and Michael S. Turner. 1993. *The Early Universe*. Reading, MA: Addison-Wesley.

Kolb, Edward W., Michael S. Turner, David Lindley, Keith Olive, and David Seckel. 1986. *Inner Space/Outer Space: The Interface between Cosmology and Particle Physics*. Chicago: University of Chicago Press.

Kragh, Helge. 1996. *Cosmology and Controversy: The Historical Development of Two Theories of the Universe*. Princeton, NJ: Princeton University Press.

Krauss, Lawrence M., and Michael S. Turner. 1995. "The Cosmological Constant Is Back." *General Relativity and Gravitation* 27: 1137–44.

Leibundgut, B., J. Spyromlio, J. Walsh, B. P. Schmidt, M. M. Phillips, N. B. Suntzeff, M. Hamuy, R. A. Schommer, R. Avilés, R. P. Kirshner, A. Riess, P. Challis, P. Garnavich, C. Stubbs, C. Hogan, A. Dressler, and R. Ciardullo. 1995. "Discovery of a Supernova (SN 1995K) at a Redshift of 0.478." *ESO Messenger* 81: 19–20.

Lightman, Alan, and Roberta Brawer. 1990. *Origins: The Lives and Worlds of Modern Cosmologists.* Cambridge, MA: Harvard University Press.

Livio, Mario, ed. 2003. *The Dark Universe: Matter, Energy, and Gravity.* Cambridge: Cambridge University Press.

Massey, Richard, et al. 2007. "Dark Matter Maps Reveal Cosmic Scaffolding." *Nature* 445: 286–90.

Mather, John C., and John Boslough. 2008. *The Very First Light: The True Inside Story of the Scientific Journey Back to the Dawn of the Universe.* New York: Basic Books.

National Research Council of the National Academies. 2003. *Connecting Quarks with the Cosmos: Eleven Science Questions for the New Century.* Washington, DC: National Academies Press.

Newberg, Heidi Jo Marvin. 1992. "Measuring $q_0$ Using Supernovae at $z \approx 0.3$." PhD diss., University of California, Berkeley, and Lawrence Berkeley Laboratory.

Newton, Isaac. 1999. *The Principia: Mathematical Principles of Natural Philosophy*, trans. I. Bernard Cohen and Anne Whitman. Berkeley: University of California Press.

O'Connor, Flannery. 1979. "Writing Short Stories." In *Mysteries and Manners.* New York: Farrar, Straus & Giroux.

Ohm, E. A. 1961. "Receiving System." *Bell System Technical Journal* 40: 1045.

Ostriker, J. P., and P. J. E. Peebles. 1973. "A Numerical Study of the Stability of Flattened Galaxies: Or, Can Cold Galaxies Survive?" *Astrophysical Journal* 186: 467–80.

Ostriker, J. P., P. J. E. Peebles, and A. Yahil. 1974. "The Size and Mass of Galaxies, and the Mass of the Universe." *Astrophysical Journal* 193: L1–L4.

Ostriker, J. P., and Paul J. Steinhardt. 1995. "The Observational Case for a Low-Density Universe with a Non-Zero Cosmological Constant." *Nature* 377: 600–2.

Overbye, Dennis. 1992. *Lonely Hearts of the Cosmos: The Scientific Quest for the Secret of the Universe.* New York: Harper Perennial.

———. 1996. "Weighing the Universe." *Science* 272: 1426–28.

Paczynski, Bohdan. 1986a. "Gravitational Microlensing at Large Optical Depth." *Astrophysical Journal* 301: 503–16.

———. 1986b. "Gravitational Microlensing by the Galactic Halo." *Astrophysical Journal* 304: 1–5.

Pais, Abraham. 1997. *'Subtle Is the Lord . . .': The Science and the Life of Albert Einstein.* Oxford: Oxford University Press.

Peacock, John A., et al. 2001. "A Measurement of the Cosmological Mass Density from Clustering in the 2dF Galaxy Redshift Survey." *Nature* 410: 169–73.

Peebles, P. J. E. 1965. "The Black-Body Radiation Content of the Universe and the Formation of Galaxies." *Astrophysical Journal* 142: 1317–26.

———. 1969. "Cosmology." *Journal of the Royal Astronomical Society of Canada* 63: 4–31.

———. 1970. "Structure of the Coma Cluster of Galaxies." *Astronomical Journal* 75: 13–20.

———. 1974. *Physical Cosmology.* Princeton, NJ: Princeton University Press.

———. 1984. "Tests of Cosmological Models Constrained by Inflation." *Astrophysical Journal* 284: 439–44.

———. 1999. "Penzias and Wilson's Discovery of the Cosmic Microwave Background." *Astrophysical Journal* 525C: 1067–68.

———. 2003. "Open Problems in Cosmology." http://arxiv.org/pdf/astro-ph/0311435v1.

———. 2009. "How I Learned Physical Cosmology." In *Finding the Big Bang*, ed. P. James E. Peebles, Lyman A. Page, Jr., and R. Bruce Partridge. New York: Cambridge University Press.

Peebles, P. J. E., and J. T. Yu. 1970. "Primeval Adiabatic Perturbation in an Expanding Universe." *Astrophysical Journal* 162: 815–36.

Peebles, P. James E., Lyman A. Page, Jr., and R. Bruce Partridge, eds. 2009. *Finding the Big Bang.* New York: Cambridge University Press.

Penzias, Arno. 1992. "The Origin of Elements." In *Nobel Lectures in Physics, 1971–1980*, ed. Stig Lundqvist. Singapore: World Scientific Publishing Co.

———. 2009. "Encountering Cosmology." In *Finding the Big Bang*, ed. P. James E. Peebles, Lyman A. Page, Jr., and R. Bruce Partridge. New York: Cambridge University Press.

Penzias, A. A., and R. W. Wilson. 1965. "A Measurement of Excess Antenna Temperature at 4080 Mc/s." *Astrophysical Journal* 142: 419–21.

Perlmutter, Saul. 1986. "An Astrometric Search for a Stellar Companion to the Sun." PhD diss., Lawrence Berkeley Laboratory and University of California, Berkeley.

Perlmutter, S., G. Aldering, M. Della Valle, S. Deustua, R. S. Ellis, S. Fabbro, A. Fruchter, G. Goldhaber, D. E. Groom, I. M. Hook, A. G. Kim, M. Y. Kim, R. A. Knop, C. Lidman, R. G. McMahon, P. Nugent, R. Pain, N. Panagia, C. R. Pennypacker, P. Ruiz-Lapuente, B. Schaefer, and N. Walton. 1998. "Discovery of a Supernova Explosion at Half the Age of the Universe." *Nature* 391: 51–54.

Perlmutter, S., G. Aldering, G. Goldhaber, R. A. Knop, P. Nugent, P. G. Castro, S. Deustua, S. Fabbro, A. Goobar, D. E. Groom, I. M. Hook, A. G. Kim, M. Y. Kim, J. C. Lee, N. J. Nunes, R. Pain, C. R. Pennypacker, R. Quimby, C. Lidman, R. S. Ellis, M. Irwin, R. G. McMahon, P. Ruiz-Lapuente, N. Walton, B. Schaefer, B. J. Boyle, A. V. Filippenko, T. Matheson, A. S. Fruchter, N. Panagia, H. J. M. Newberg, and W. J. Couch. 1999. "Measurements of Ω and Λ from 42 High-Redshift Supernovae." *Astrophysical Journal* 517: 565–86.

Perlmutter, S., S. Deustua, S. Gabi, G. Goldhaber, D. Groom, I. Hook, A. Kim, M. Kim, J. Lee, R. Pain, C. Penypacker, I. Small, A. Goobar, R. Ellis, R. McMahon, B. Boyle, P. Bunclark, D. Carter, K. Glazebrook, M. Irwin, H. Newberg, A. V. Filippenko, T. Matheson, M. Dopita, J. Mould, and W. Couch. 1995. "Scheduled Discoveries of 7+ High-Redshift Supernovae: First Cosmology Results and Bounds on $q_0$." Presented at the NATO Advanced Study Institute Thermonuclear Supernovae Conference, Aiguablava, Spain, June 20–30, 1995b.

Perlmutter, S., S. Gabi, G. Goldhaber, A. Goobar, D. E. Groom, I. M. Hook, A. G. Kim, M. Y. Kim, J. C. Lee, R. Pain, C. R. Pennypacker, I. A. Small, R. S. Ellis, R. G. McMahon, B. J. Boyle, P. S. Bunclark, D. Carter, M. J. Irwin, K. Glazebrook, H. J. M. Newberg, A. V. Filippenko, T. Matheson, M. Dopita, and W. J. Couch. 1997. "Measurements of the Cosmological Parameters Ω and Λ from the First Seven Supernovae at $z \geq 0.35$." *Astrophysical Journal* 483: 565–81.

Perlmutter, S., C. R. Pennypacker, G. Goldhaber, A. Goobar, R. A. Muller, H. J. M. Newberg, J. Desai, A. G. Kim, M. Y. Kim, I. A. Small, B. J. Boyle, C. S. Crawford, R. G. McMahon, P. S. Bunclark, D. Carter, M. J. Irwin, R. J. Terlevich, R. S. Ellis, K. Glazebrook, W. J. Couch, J. R. Mould, T. A. Small, and R. G. Abraham. 1995a. "A Supernova at $z = 0.458$ and Implications for Measuring the Cosmological Deceleration." *Astrophysical Journal Letters* 440: L41–L44.

Petrossian, V., E. Salpeter, and P. Szekeres. 1967. "Quasi-Stellar Objects in Universes with Non-Zero Cosmological Constant." *Astrophysical Journal* 147: 1222–26.

Phillips, M. M. 1993. "The Absolute Magnitudes of Type Ia Supernovae." *Astrophysical Journal* 413: L105–L108.

Raup, David M., and J. John Sepkoski, Jr. 1984. "Periodicity of Extinctions in the Geologic Past." *Proceedings of the National Academy of Science USA* 81: 801–5.

Riess, Adam G., William H. Press, and Robert P. Kirshner. 1995a. "Using Type Ia Supernova Light Curve Shapes to Measure the Hubble Constant." *Astrophysical Journal Letters* 438: L17–L20.

———. 1995b. "Determining the Motion of the Local Group Using SN Ia Light Curve Shapes." *Astrophysical Journal Letters* 445: L91–L94.

———. "A Precise Distance Indicator: Type Ia Supernova Multicolor Light-Curve Shapes." *Astrophysical Journal* 473: 88–109.

Riess, Adam G., Alexei V. Filippenko, Peter Challis, Alejandro Clocchiatti, Alan Diercks, Peter M. Garnavich, Ron L. Gilliland, Craig J. Hogan, Saurabh Jha, Robert P. Kirshner, B. Leibundgut, M. M. Phillips, David Reiss, Brian P. Schmidt, Robert A. Schommer, R. Chris Smith, J. Spyromilio, Christopher Stubbs, Nicholas B. Suntzeff, and John Tonry. 1998. "Observational Evidence from Supernovae for an Accelerating Universe and a Cosmological Constant." *Astronomical Journal* 16: 1009–38.

Riess, Adam G., Peter E. Nugent, Ronald L. Gilliland, Brian P. Schmidt, John Tonry, Mark Dickinson, Rodger I. Thompson, Tamás Budavári, Stefano Casertano, Aaron S. Evans, Alexei V. Filippenko, Mario Livio, David B. Sanders, Alice E. Shapley, Hyron Spinrad, Charles C.

Steidel, Daniel Stern, Jason Surace, and Sylvain Veilleux. 2001. "The Farthest Known Supernova: Support for an Accelerating Universe and a Glimpse of the Epoch of Deceleration." *Astrophysical Journal* 560: 49–71.

Riess, Adam G., Louis-Gregory Strolger, John Tonry, Stefano Casertano, Henry C. Ferguson, Bahram Mobasher, Peter Challis, Alexei V. Filippenko, Saurabh Jha, Weidong Li, Ryan Chornock, Robert P. Kirshner, Bruno Leibundgut, Mark Dickinson, Mario Livio, Mauro Giavalisco, Charles C. Steidel, Txitxo Benítez, and Zlatan Tsvetanov. 2004. "Type Ia Supernova Discoveries at $z > 1$ from the Hubble Space Telescope: Evidence for Past Deceleration and Constraints on Dark Energy Evolution." *Astrophysical Journal* 607: 665–87.

Riess, Adam G., Louis-Gregory Strolger, Stefano Casertano, Henry C. Ferguson, Bahram Mobasher, Ben Gold, Peter J. Challis, Alexei V. Filippenko, Saurabh Jha, Weidong Li, John Tonry, Ryan Foley, Robert P. Kirshner, Mark Dickinson, Emily MacDonald, Daniel Eisenstein, Mario Livio, Josh Younger, Chun Xu, Tomas Dahlén, and Daniel Stern. 2007. "New Hubble Space Telescope Discoveries of Type Ia Supernovae at $z >= 1$: Narrowing Constraints on the Early Behavior of Dark Energy." *Astrophysical Journal* 659: 98–121.

Riordan, Michael, and David N. Schramm. 1991. *The Shadows of Creation: Dark Matter and the Structure of the Universe.* New York: W. H. Freeman.

Rubin, Vera Cooper. 1951. "Differential Rotation of the Inner Metagalaxy." *Astronomical Journal* 1190: 47–48.

———. 1954. "Fluctuations in the Space Distribution of the Galaxies." *Proceedings of the National Academy of Sciences* 40: 541–49.

———. 1983. "The Rotation of Spiral Galaxies." *Science* 220: 1339–44.

———. 1997. *Bright Galaxies, Dark Matters.* Woodbury, NY: AIP Press.

———. 2003. "A Brief History of Dark Matter." In *The Dark Universe: Matter, Energy, and Gravity*, ed. Mario Livio. Cambridge: Cambridge University Press.

———. 2006. "Seeing Dark Matter in the Andromeda Galaxy." *Physics Today* 59: 8–9.

Rubin, Vera C., Jaylee Burley, Ahmad Kiasatpoor, Benny Klock, Gerald Pease, Erich Rutscheidt, and Clayton Smith. 1962. "Kinematic Studies of Early-Type Stars, I: Photometric Survey, Space Motions, and Comparison with Radio Observations." *Astronomical Journal* 67: 491–531.

Rubin, Vera C., W. Kent Ford, Jr., and Judith S. Rubin. 1973. "A Curious Distribution of Radial Velocities of Sc I Galaxies with $14.0 \leq m \leq 15.0$." *Astrophysical Journal* 183: L111–L115.

Rubin, Vera C., W. Kent Ford, Jr., Norbert Thonnard, Morton S. Roberts, and John A. Graham. 1976a. "Motion of the Galaxy and the Local Group Determined from the Velocity Anisotropy of Distant Sc I Galaxies, I: The Data." *Astronomical Journal* 81: 687–718.

Rubin, Vera C., Norbert Thonnard, W. Kent Ford, Jr., and Morton S. Roberts. 1976b. "Motion of the Galaxy and the Local Group Determined from the Velocity Anisotropy of Distant Sc I Galaxies, II: The Analysis for the Motion." *Astronomical Journal* 81: 719–37.

Rubin, Vera C., W. Kent Ford, Jr., and Norbert Thonnard. 1978. "Extended Rotation Curves of High-Luminosity Spiral Galaxies, IV: Systematic Dynamical Properties, SA→SC." *Astrophysical Journal* 225: L107–L111.

Sandage, Allan. 1970. "Cosmology: The Search for Two Numbers." *Physics Today* 23: 34–41.

———. 1987. "Observational Cosmology 1920–1985: An Introduction to the Conference." In *Observational Cosmology* (Proceedings of the 124th Symposium of the International Astronomical Union, Beijing, China, Aug. 25–30, 1986), ed. Adelaide Hewitt, Geoffrey Burbidge, and Li Zhi Fang. Dordrecht, Neth.: D. Reidel.

Smith, Sinclair. 1936. "The Mass of the Virgo Cluster." *Astrophysical Journal* 83: 23–30.

Smoot, George, and Keay Davidson. 1994. *Wrinkles in Time: Witness to the Birth of the Universe.* New York: Harper Perennial.

Stachel, John, ed. 1998. *Einstein's Miraculous Year: Five Papers That Changed the Face of Physics.* Princeton: Princeton University Press.

Staniszewski, Z., et al. 2009. "Galaxy Clusters Discovered with a Sunyaev-Zel'dovich Effect Survey." *Astrophysical Journal* 701: 32–41.

Thompson, Silvanus P. 1910. *The Life of William Thomson, Baron Kelvin of Largs.* London: Macmillan and Co.

Tolman, Richard C. 1987. *Relativity, Thermodynamics, and Cosmology.* New York: Dover.

Tryon, Edward P. 1973. "Is the Universe a Quantum Fluctuation?" *Nature* 246: 396–97.

Turner, Kenneth C. 2009. "Spreading the Word—or How the News Went from Princeton to Holmdel." In *Finding the Big Bang*, ed. P. James E. Peebles, Lyman A. Page, Jr., and R. Bruce Partridge. New York: Cambridge University Press.

Turner, Michael S. 1998a. "Cosmology Solved? Maybe." http://arxiv.org/abs/astro-ph/9811366.

———. 1998b. "Cosmology Solved?" http://arxiv.org/abs/astro-ph/9811447.

———. 1999. "Cosmology Solved? Quite Possibly!" *Publications of the Astronomical Society of the Pacific* 111: 264–73.

———. 2009. "David Norman Schramm, 1945–1997: A Biographical Memoir." Washington, DC: National Academy of Sciences.

Turner, M. S., G. Steigman, and L. M. Krauss. 1984. "Flatness of the Universe: Reconciling Theoretical Prejudices with Observational Data." *Physical Review Letters* 52: 2090–93.

Uomoto, A., and R. P. Kirshner. 1985. "Peculiar Type I Supernovas." *Astronomy and Astrophysics* 149: L7–L9.

van Bibber, Karl, and Leslie J. Rosenberg. 2006. "Ultrasensitive Searches for the Axion." *Physics Today* 59 (Aug.): 30–35.

Weinberg, Steven. 1993. *The First Three Minutes: A Modern View of the Origin of the Universe.* New York: Basic Books.

White, Simon D. M. 2007. "Fundamentalist Physics: Why Dark Energy Is Bad for Astronomy." *Reports on Progress in Physics* 70: 883–97.

Wilczek, Frank. 1985. "Conference Summary and Concluding Remarks." In *The Very Early Universe,* ed. G. W. Gibbons, S. W. Hawking, and S. T. C. Siklos. Cambridge: Cambridge University Press.

———. 1991. "The Birth of Axions." *Current Contents* 22: 8–9.

Wilkinson, David T. 2009. "Measuring the Cosmic Microwave Background Radiation." In *Finding the Big Bang*, ed. P. James E. Peebles, Lyman A. Page, Jr., and R. Bruce Partridge. New York: Cambridge University Press.

Wilson, Robert W. 1992. "The Cosmic Microwave Background Radiation." In *Nobel Lectures in Physics, 1971–1980*, ed. Stig Lundqvist. Singapore: World Scientific Publishing Co.

———. 2009. "Two Astronomical Discoveries." In *Finding the Big Bang*, ed. P. James E. Peebles, Lyman A. Page, Jr., and R. Bruce Partridge. New York: Cambridge University Press.

Zwicky, Fritz. 1933. "Die Rotverschiebung von extragalaktischen Nebeln." *Helvetica Physica Acta* 6: 110–27.

———. 1937. "On the Masses of Nebulae and of Clusters of Nebulae." *Astrophysical Journal* 86: 217–46.

# Index

## About the Author

Richard Panek is the author of the critically acclaimed books *The Invisible Century: Einstein, Freud, and the Search for Hidden Universes* and *Seeing and Believing: How the Telescope Opened Our Eyes and Our Minds to the Heavens.* A recipient of a John Simon Guggenheim Memorial Foundation fellowship, he writes on science and culture for the *New York Times, Discover, Natural History, Esquire,* and *Seed,* among many other publications.